# Weathered

## Cultures of Climate

In memory of my mother, Shelagh Mary Hulme, who died as I embarked on the writing of this book. Her love of geography and of literature are reflected in various ways in what is written here.

# Weathered
## Cultures of Climate
Mike Hulme

Los Angeles | London | New Delhi
Singapore | Washington DC | Melbourne

Los Angeles | London | New Delhi
Singapore | Washington DC | Melbourne

Editor: Robert Rojek
Editorial assistant: Matthew Oldfield
Production editor: Katherine Haw
Copyeditor: Catja Pafort
Proofreader: Camille Bramall
Indexer: Bill Johncocks
Marketing manager: Sally Ransom
Cover design: Stephanie Guyaz
Typeset by: C&M Digitals (P) Ltd, Chennai, India
Printed and bound by CPI Group (UK) Ltd,
Croydon, CR0 4YY

**Library of Congress Control Number: 2016939661**

**British Library Cataloguing in Publication data**

A catalogue record for this book is available from
the British Library

ISBN 978-1-4739-2498-7
ISBN 978-1-4739-2499-4 (pbk)

# Praise for *Weathered*

'Mike Hulme's wise and well-crafted book encircles the idea of climate from a series of perspectives, showing its elusive nature from a welter of examples. As the argument develops, we see how climate is embedded in multiple cultures, histories, and knowledges about nature. We are shown how our views of climate depend on personal experiences, scientific models, inherited tropes, and political interest. Each chapter reflects a turn of the kaleidoscope, gradually making the reader see both the complexity and the singularity of each image. Hulme's remarkable achievement is to humanise climate, without losing sight of the larger challenges; this is where the book cannot but affect the reader.'

**Kirsten Hastrup, Professor of Anthropology, University of Copenhagen**

'Everybody may be talking about the weather, but how do we experience climate? While climate has mostly been left to the natural sciences, Mike Hulme's book shows how climate is much more than the "average weather". It is a cultural relationship between humans and the weathers they dwell in. How do cultures live with the weather? How does the experience of climate structure our sense of space and time? This book is the first to offer a systematic overview of the many forms of knowledge, cultural practices and personal attitudes that helped humans in different epochs and locations deal with their meteorological environment. Its importance lies not just in the wealth of material and its brilliantly clear structure but also in the way Hulme links a humanities-based approach to climate with the current state of climate science. This is a milestone for interdisciplinary climate research and a must-read for all scholars and students trying to understand how a human being-in-the-world is a being-in-climate.'

**Eva Horn, Faculty of Philological and Cultural Studies, University of Vienna**

'We desperately need a book like this one, a book that reorients our thinking about climate change from temperature and precipitation to culture, values, emotions, and social justice. Mike Hulme has delivered beautifully in this highly accessible, boldly insightful, and elegant book. *Weathered* divulges quite clearly the complex ways we think about weather and climate. And it also shows us that when we define or explain, study or represent, fear or blame, engineer or predict the climate, we are ultimately empowering some people while disempowering others. Anyone who cares about climate – from climate scientists and policymakers to journalists, geographers, historians, students, and activists – should read this book.'

**Mark Carey, Author of** *In the Shadow of Melting Glaciers: Climate Change and Andean Society*

'In his bracing new book, Mike Hulme throws open cultural windows on climate, illuminating its history and geography as a powerful form of human experience and imagination. Through a series of frameworks, concerning knowledge, narrative, livelihood and policy, and a rich range of examples, from scientific modelling to impressionist painting, statistical mapping to song and dance, Hulme guides his readers, clearly and accessibly, through the cultural worlds of climate. *Weathered* introduces students from many subjects to the many meanings and functions of climate, and its relations to such matters as commerce and creativity. The book also challenges scholars in many fields of science and the humanities to see beyond their specialisms, in such a pressing field of inquiry and concern.'

**Stephen Daniels, Professor of Cultural Geography, University of Nottingham**

# Contents

# List of Boxes

# List of Figures

# About the Author

Mike Hulme is professor of climate and culture in the Department of Geography at King's College London. His work sits at the intersection of climate, history and culture. He studies how knowledge about climate and its changes is made and represented, and analyses the numerous ways in which the idea of climate-change is deployed in public discourse around the world.

# Preface

This book is about the idea of climate, an enduring idea of the human mind and also a powerful one. Today, the idea of climate is most commonly associated with the phenomenon and discourse of climate-change[1] and its scientific, economic, religious, ethical, social and political dimensions. I have written about these axes of public argumentation in an earlier book – *Why We Disagree About Climate Change* (Hulme, 2009) – but before the cultural politics of climate-change can truly be understood, I believe a richer understanding of the idea of climate itself is needed. Because climate-change is such a pervasive phenomenon and discourse which is re-making the contemporary world, it is important to take a step back and undertake historical, geographical and cultural investigations of the idea of climate itself.

Like any interesting word, 'climate' defies easy definition for reasons explained in **Chapter 1**. My argument in *Weathered* is that climate – as it is imagined, studied and acted upon – needs to be understood, first and foremost, culturally. Since climate is a complex and abstract idea, it cannot be understood independently of the cultures within which the idea takes shape. This argument challenges the primacy of natural science definitions of climate and, hence, questions the predominantly scientific understanding of (human-caused) climate-change. For example, the successive assessment reports of the Intergovernmental Panel on Climate Change (the IPCC) promote the view that climate should be understood as a planetary system of physically interconnecting processes, a system which can adequately be represented and hence predicted using mathematical models. This framing of climate is dominant in much academic scholarship, in politics and in public debates. It assumes that changes in climate, and in its human and non-human drivers, are to be studied, explained and predicted through scientific theory and observation. As a consequence, forensic detection and attribution studies using ensembles of complex climate models and arcane statistics have become central to the scientific and public status of the reality of climate-change.

But there is another story to be told about climate-change, one which starts with the cultural origins of the idea of climate. Rather than framing climate solely as an interconnected global physical system or as a statistical artefact of

---

[1] Throughout this book I use the construction 'climate-change' to refer to the contemporary idea of human-caused global climatic change. In this way I differentiate the physical and discursive realities of anthropogenic changes in global climate from other expressions of change, for example, 'climate change' (un-hyphenated), 'changes in climate' or 'climatic change'.

repeated weather measurements, in this book I develop the case that climate needs also to be understood as an idea that is given meaning through cultures. This means that climate pre-eminently can be – and indeed has been – changed by cultures. If this is true then climate has a cultural geography and history which is interwoven with its physical geography and history. This geography and history of climate forms the substrate out of which contemporary beliefs, claims and disputes about climate-change emerge today. *Weathered* therefore places the contemporary phenomenon and discourse of climate-change within a matrix of rich cultural understandings and meanings.

My approach to understanding climate extends well beyond the traditional disciplines where climates and cultures are studied – climatology and anthropology, respectively – even if within these disciplines there are identifiable communities of scholars concerned with their interactions. For example, Thornes and MacGregor (2003) identify a cultural climatology tradition within geography, while anthropologist Todd Sanders refers to a circle of 'critical climate change anthropologists' (Sanders, 2014). To fully grasp the idea of climate – and, by implication, I would argue, the phenomenon and discourse of climate-change – the insights of geographers, anthropologists, historians, psychologists, sociologists, theologians, historians, eco-critics and philosophers are needed. And if science is correctly understood as a cultural pursuit, then insights from sociologists of science and scholars of science and technology studies are also needed.

*Weathered: Cultures of Climate* therefore opens up the many ways in which the idea of climate is given meaning in different human cultures and how it is used; how climates are historicised, known, changed, lived with, blamed, feared, represented, predicted, governed and, at least putatively, redesigned. These actions performed on or with the idea of climate emerge from the diverse cultural interpretations of humans' sensory experience of the atmosphere's restless weather. *Weathered* develops a case for understanding climate as an enduring, yet malleable, idea which humans use to stabilise cultural relationships with their weather. The discursive phenomenon of climate-change should therefore be understood as a ubiquitous trope through which the material, psychological and cultural agency of climate is exercised in today's world. In this sense the phenomenon of climate-change is not a decisive break from the past, neither is it a unique outcome of modernity. Climate-change should rather be seen as the latest stage in the cultural evolution of the idea of climate, an idea which enables humans to live with their weather through a widening and changing range of cultural and material artefacts, practices, rituals and symbols.

* * *

The interactions between weather and culture appear everywhere in daily life; for example, in social memories of past climatic extremes, in emotional moods, in technologies of adaptation, in fiction, poetry and song, in narratives of blame, in dress codes and so on. Many of these relationships have been written about, but in disparate texts and journals, fragmented across many different academic disciplines. A coherent literature which treats the rich interactions between climates and cultures in a systematic way is lacking. The number and scope of monographs and reference texts offering a synoptic view of climate and culture is rather limited, and some of them obscure. Monographs by Boia (2005), Behringer (2010), Leduc (2010; 2016), Bristow and Ford (2016) and, in Japanese, Watsuji (1988[1935]) are noteworthy exceptions, as are the edited collections of Strauss and Orlove (2003), Crate and Nuttall (2009), Dove (2014), Barnes and Dove (2015) and, in German, Welzer et al. (2010). But this deficiency is slowly being remedied. As the new sub-field of environmental humanities expands, and as the provocative idea of the Anthropocene continues to irrupt in intellectual and cultural worlds, the next few years will see a growing number of texts which open up new ways for thinking about the relationships between climates and cultures.

I have already given some visibility and structure to this growing scholarship in a recently published six-volume SAGE Major Reference Work, *Climates and Cultures* (Hulme, 2015a). That collection of 88 published articles and book chapters from a range of disciplines captures and organises some of the most important academic writing on climate and culture that has appeared since the early 1990s. It provides a series of signposts to guide readers through a growing body of work on climates and cultures from the disciplines of human geography, environmental history, philosophy, science studies, anthropology, eco-criticism, sociology and religious studies. I organised that collection into six themes:

Vol. 1    Cultures of Climate Knowledge

Vol. 2    Historical Readings of Climate

Vol. 3    Climate and Agency

Vol. 4    Climate and Culture in Places and Practices

Vol. 5    Cultural Readings of Future Climate

Vol. 6    Climate Change in Literary, Visual and Performance Cultures

*Weathered: Cultures of Climate* should be seen as a companion text to that earlier reference work, offering an extended synthesis of the published work on climates and cultures. Following an opening definitional chapter, each of the following 11 chapters in this book explores a different aspect of how cultures use, or have used, the idea of climate to make sense of their relationship with the weather. The 11 themes I have selected are certainly not exhaustive and the distinctions between some of them may blur. Yet taken together they illustrate the cultural foundations of human knowledge of climate (Part 1), human life with climate (Part 2) and human relationships with future climates (Part 3).

A word is needed about the headline title of the book, *Weathered*. Like climate, weather is a rich word in the English language. It can be used as noun, verb and adjective. As noun, weather refers to the instantaneous physical state of the atmosphere; hence there are many types of weather to be described: stormy, calm, lovely, threatening, changeable, benign and so on. As verb, the process of weathering refers to the exposure of objects or sentient beings to various types of weather; for example, 'these buildings are weathering well'. This exposure, usually over significant periods of time, changes the object or being in characteristic and often permanent ways so that it might be described as weathered: it bears in some way the imprint of the weather to which it has been exposed. In this third case then, weather becomes an adjective, as in 'her weathered face betrayed years of exposure to the wind and sun of the mountain climate'.

I am using the term 'weathered' in this book in this latter sense. As I explain in Chapter 1, climate is an idea that helps stabilise the human experience of weather and allows humans to live culturally with their weather. Cultures and individuals can be thought of as being 'weathered' through repeated exposures over periods of time to particular types and sequences of weather. It is reflecting on this process of weathering which, I suggest, helps appreciate the value of the idea of climate. Cultures bear the imprint of the weather in which they exist and to which they respond. As with trees or human bodies, cultures too are weathering; indeed, they cannot avoid being weathered in some way or other. It is these interactions between cultures and weather that I seek to explore in this book and which, I will argue, offer a richer and more intimate understanding of the idea of climate than can be offered by science alone. As environmental historian Alexander Wilson explains in relation to landscape, 'In the broadest sense of the term, landscape is a way of seeing the world and imagining our relationship to nature. It is something we think, do and make as a social collective' (Wilson, 1992: 17). Replace 'landscape' and 'nature' with 'climate' and 'weather' in the above description and you have a good summary of the argument of this book.

Mike Hulme
King's College London
April 2016

# Acknowledgements

This book was written over the winter of 2015/16, although its origins date back to my stay at the Rachel Carson Centre at the Ludwig Maximilian University (LMU) in Munich during the summer of 2014. I would like to thank the Centre and its directors – Christof Mauch and Helmuth Trischler – for the award of a Carson Writing Fellowship that facilitated this visit and also the many colleagues at the Centre with whom I interacted and was able to share my thinking. I would like additionally to thank George Adamson, Georgina Endfield, Greg Garrard, Jim Fleming, Kirsten Hastrup, Kjersti Fløttum, Mathias Heymann, Edvard Hviding, Vlad Janković, Willis Jenkins, Adeline Johns-Putra, Myanna Lahsen, Eva Lövbrand, Amanda Machin, Martin Mahony, Ruth Morgan, Kate Porter, Sam Randalls, Steve Rayner, Peter Rudiak-Gould, Birgit Schneider, Joe Smith and Chaya Vaddhanaphuti for stimulating conversations over recent years through which some of these ideas in this book developed and evolved. I am also indebted to Gill Hulme and Emma Hulme for putting me straight on Tolkien's Middle Earth and Game of Thrones, respectively. George Adamson, Eliza de Vet, Martin Mahony, Lucy Rose and Dylan Robinson each read through one or more draft chapters and their close reading led to some significant improvements in the text, although the responsibility for what finally appears in print remains fully mine.

Some of the ideas in this book were first articulated in earlier published work and so I should acknowledge that Chapter 1 is partly based on my essay published in *GEO: Geography and Environment* (Hulme, 2015b), Chapter 7 draws upon an article in *The Geographical Journal* (Hulme, 2008) and parts of Chapter 10 upon my book *Can Science Fix Climate Change?* (Hulme, 2014a) and an essay in *Current Anthropology* (Hulme, 2015c).

At SAGE I would like to thank Robert Rojek for commissioning the book and Matt Oldfield, Catja Pafort for her copyediting work, and Katherine Haw in the production team. SAGE commissioned four anonymous readers of the original book proposal and they each made valuable suggestions which improved the framing of various chapters. I also thank Bill Johncocks who has produced, yet again, an index of the highest professional standard.

Finally, my greatest debt is to my wife, Gill, who graciously accepted the intrusion of this book into numerous Norwich weekends and an entire walking holiday in Derbyshire. Thank you for your love, support and patience.

# 1

# What is Climate?

## Introduction

This needs saying right from the start: climate is hard to place and even its existence is questionable. It seems to be everywhere (Can you escape from climate? Is anywhere on Earth climate-less?) and yet it is nowhere (Can you point to climate or take me to see it?). People seem to know intuitively what climate is and yet they struggle to articulate an adequate definition of it. And yet if climates *didn't* exist they would have to be invented; in fact, maybe they are invented. The value of the idea of something *like* 'climate' is evidenced by the wide metaphorical usage of it in everyday speech, referencing different intangible realities. Thus, political, intellectual, moral and economic climates are readily invoked. Inventing ideas is of course what people have done throughout history and what they continue to do across diverse cultures. And the idea of climate is nearly as old as recorded human history and as ubiquitous as, well, the weather. Climate appears to be a necessary invention if people are to make sense of the world in which they live.

So what exactly *is* climate? Is it merely an idea and, if so, what sort of idea? We might say that climatologists are able to conjure climates into statistical existence through averaging meteorological measurements made repeatedly at the same place day after day. Climates therefore exist abstractly as numbers. We might also recognise that Earth system scientists are able to simulate climates into virtual existence inside their computers, reproducing *in silico* the workings of a physically connected global Earth system and generating terrabytes of 'climate data'. Climates therefore also exist virtually as numerical models. And then geographers are able to take you to different physical places on Earth and show you the effects of different climates on landscapes (the Sahara or the tundra), on ecosystems (alpine or tropical ) or on the design of buildings which humans inhabit (Swiss chalets or Balinese houses).

But even though we live *in* climates I cannot *show* you climate. I can place you in the middle of a hurricane and we can shiver together in a blizzard or be awe-struck by observing a tornado. We might admire the beauty of a tranquil sunset or feel the reverberations of thunder and tremble in awe. But hurricanes, blizzards and thunderstorms are merely transient weather events, the outworking of a restless and constantly changing atmosphere. Climate is something else, hinting at a physical reality that is both more stable and durable than the weather. Unlike the weather, climate is therefore an idea of the human mind.

\* \* \*

In this opening chapter I explore the idea of climate and introduce the scope of the chapters that follow. I propose that climate best be understood as an idea which mediates the sensory experience of ephemeral weather and the cultural ways of living which humans have developed to accommodate this experience. The idea of climate connects material and imaginative worlds in ways that create order and offer stability to human existence. People could not live without their climate.

## Definitions of Climate

The operational scientific definition of climate usually starts with something like the official wording used by the World Meteorological Organisation (WMO): climate is '… a statistical description in terms of the mean and variability of relevant quantities of certain variables (such as temperature, precipitation or wind) over a period of time ranging from months to thousands or millions of years' (WMO, n.d.). This description conventionally relies on 30 years of weather data. Climate might also be understood in a more general scientific sense as a description of the state and dynamics of the physical planetary system, which consists of 'five major components: the atmosphere, the hydrosphere, the cryosphere, the lithosphere and the biosphere, and the [evolving] interactions between them' (IPCC, 2013: 1451). This would be the way in which climate modellers would seek to understand and simulate climate in their computers.

Such definitions do not do justice to the deep material and symbolic interactions which occur between weather and cultures in places; interactions which I argue in this book are central to the idea of climate. They too easily maintain a false separation between a physical world (to be understood through scientific inquiry) and an imaginative one (to be understood through meaningful narratives or human rituals). Such a distinction maps easily onto the nature–culture dualism which has engrained itself in much western thought and practice, but which has been subject to extensive scholarly deconstruction over recent decades (e.g. Latour, 1993; Plumwood, 1993; Castree, 2013) (see **Box 1.1**).

---

### Box 1.1: Climate and *Mejatoto*

The western separation of nature from culture is far from ubiquitous in today's world. This has significance when seeking to understand how the idea of climate works in different cultures. Anthropologist Peter Rudiak-Gould's work in the Marshall Islands of the western Pacific helps here. In Marshallese, the closest equivalent to the English word climate is *mejatoto*. Yet *mejatoto* means more than just 'weather', 'air' or 'climate'. It refers to a much wider range of attributes of the dwelt-in environment, people's physical as well as their cultural surroundings. When asked by Rudiak-Gould whether the *mejatoto* had changed, Marshallese interviewees replied, '"Yes. People are not like they were in the past. The culture has changed". "Yes. People are lazy now". "Yes. The *mejatoto* is bad now – people do not cooperate with their families like they used to". "Yes. Life has changed. They do not take care of each other like they used to"' (Rudiak-Gould, 2012: 49). When Marshall Islanders talk about their 'climate', they are talking about the seamless physical and social surroundings in which they live and through which their lives make sense. *Mejatoto* refers to the matrix of material and cultural relationships which bring order and meaning to their lives. A change in the *mejatoto* – *oktak in mejatoto* – is therefore an unsettling of all such relationships. As Rudiak-Gould observes, '*Mejatoto* is not polysemous [possessing multiple meanings] per se – it only appears that way to those from a cultural background that separates nature and culture' (2012: 50).

## Experiences of Weather

Contrary to such a dualist position, this book proposes a different way of approaching the idea of climate which requires thinking more directly about the weather and what it means to people. A standard dictionary definition of weather would be '... a description of the state of the atmosphere with respect to wind, temperature, cloudiness, moisture, pressure and so on'. It is such instantaneous meteorological conditions which, measured scientifically and then averaged over a period of time, generate the conventional statistical definition of climate offered above. It is a definition that dates back no more than two centuries and was only made possible by the invention of meteorological instruments in the seventeenth century.

But climate is not weather. Weather has an immediacy and evanescence that climate does not have. Weather is in flux; it is always both passing away and being renewed. Weather captures the instantaneous atmospheric conditions in which sentient creatures live, sense, imagine and build. Weather can be seen, heard and felt, as expressed in this passage from the Japanese philosopher Tetsuro Watsuri,

> A cold wind may be experienced as a mountain blast or the cold dry wind that sweeps through Tokyo at the end of winter. The spring breeze may be one which blows off cherry blossoms or which caresses the waves... As we find our gladdened or pained selves in a wind that scatters the cherry blossoms, so do we apprehend our wilting selves in the very heat of summer that scorches down on plants and trees in a spell of dry weather. (Watsuri, 1988[1935]: 5)

It is this sensory experience of weather, and its material and emotional effects, that conditions a diverse array of cultural responses to human dwelling in the atmosphere; for example, celebratory rituals, material technologies, cultural memories and social practices. Clothes are designed to withstand cold and buildings to withstand wind; the coming of the cherry blossom and the onset of the monsoon are celebrated; weather prophets are designated to forecast the future state of the atmosphere. These cultural artefacts, moods and practices, inspired by diverse experiences of weather – some benign, some threatening – give shape and meaning to human lives. They are what Eliza de Vet (2013; 2014) calls 'weather ways': the variations that occur between repeated practices as individuals and communities adjust culturally to the weather. People live with their weather culturally; indeed, there is no way to live with weather other than culturally.

## The Idea of Climate

Beyond the concepts and definitions offered by the WMO and climate scientists, I suggest climate is better understood as an idea which mediates between the human experience of ephemeral weather and the cultural ways of living which are animated by this experience. The idea of climate introduces a sense of stability or normality into what otherwise would be too chaotic and disturbing an experience of unruly and unpredictable weather. The weather humans experience often fails to meet their expectations. But the fact that people do *have* expectations is due to the idea of climate, as geographer and climatologist Kenneth Hare recognised years ago: 'Climate is the ordinary man's [sic] expectation of weather ... there is a limit to the indignities that the weather can put upon him, and he can predict what clothes he will need for each month of the year' (Hare, 1966: 99–100). Holding on to climate as a normalising idea offers humans a certain sense of security; it allows them to 'put weather in its place' so to speak (see **Box 1.2**). Or as historian of science Lorraine Daston explains in her essay exploring the boundaries of nature, '... without well-founded expectations, the world of causes and promises falls apart' (Daston, 2010: 32).

---

### Box 1.2: English Weather-Talk

In her best-selling book from 2004, *Watching the English: the Hidden Rules of English Behaviour*, anthropologist Kate Fox devoted a whole chapter to the idea of weather-talk. Drawing upon Dr Johnson's famous aphorism from the eighteenth century that 'When two Englishmen meet, their first talk is of the weather', Fox helpfully explains the nature of such weather talk. These conversation-starters are often expressed as rhetorical questions: 'Ooh, isn't it cold today?' or 'Isn't it a nice day?' English weather-talk is

---

therefore a ritualistic form of social interaction which helps overcome reticence or unfamiliarity. But the shared idea of climate is the unspoken context which allows this weather-talk to operate as a communicative form of social grooming. Most such pleasantries refer to the shared expectation of what the weather *should be* like today: 'It's mild for February, isn't it?' or 'Yesterday's storm was so unseasonable'. Climate is the reassuring idea that there is regularity and a normality to which ultimately the weather will conform, even if today it is not doing so. Such weather-talk is an expression of solidarity in our shared sense of a secure climate. So although the English might not mention climate *explicitly* in their weather-talk, one can see how the idea of an expected English climate is the shared tacit framework which brings intelligibility, order and meaning into their social interactions.

If, as phenomenologist Julien Knebusch explains, '... climate refers to a cultural relationship established progressively between human beings and weather' (Knebusch, 2008: 246), the idea of climate should be understood as performing important psychological and cultural functions. Climate offers a way of navigating between the human experience of a constantly changing atmosphere, with its attendant insecurities, and the need to live with a promise of stability and regularity. This is what Nico Stehr refers to as 'trust in climate' (Stehr, 1997). Climate offers an ordered container, a linguistic, numerical or sensory repertoire, through which the unsettling arbitrariness of the restless weather is interpreted and tamed. This container creates Daston's necessary orderliness. The idea of climate helps stop the world falling apart. This is one of the reasons why the idea of climate *changing* is so unsettling: it undermines the 'trust' people place in climate as a cultural symbol of large-scale orderliness, an invention which eases their anxieties about the weather. Novelist Margaret Atwood captures this unease when discussing the idea of a changing climate: 'I think calling it climate change is rather limiting. I would rather call it the *everything change*. Everything is changing in ways that we cannot yet fully understand or predict' (quoted in Romm, 2015).

As mentioned earlier, climate *may* be defined according to the aggregated statistics of weather in places (the WMO) or as a scientific description of an interacting physical system (the IPCC). These definitions have value in allowing scientists to study the physicality of weather and climate. But if climate is understood as no more than this, then something crucial is being missed; namely, how the idea of climate emerges from the innumerable ways in which weather and cultures are mutually shaping and changing each other. It is these interactions that I suggest are captured by the idea of climate, an idea which therefore functions to stabilise cultural relationships between people and their weather. Beyond scientific analysis, the full richness of the idea climate can only be acquired imaginatively, as an idea of what the weather of a place 'should be' at a certain time of year and held in social or personal memory and given diverse cultural expression.

But however defined, whether scientifically or culturally, it is the human sense of climate that establishes certain expectations about the atmosphere's performance and how we respond to it. The idea of climate cultivates the possibility of a stable psychological life and of meaningful human action in the world. Put simply, the idea of climate allows humans to live culturally with their weather. It is in this sense then that I offer the idea of cultures as being 'weathered'. Furthermore, I argue that geographical and historical investigations of this weathering process yield deeper and richer understandings of the idea of climate than can be issued by natural scientists. It is why the account of climate offered in *Weathered* might be read as a cultural geography of climate (see **Box 1.3**).

---

### Box 1.3: Culture and Cultural Geography

As with climate, culture is a rich word with multiple and complex meanings. At one level one might think of culture simply as 'what humans do', for example, an aggregation of the artefacts, ideas, practices, symbols and emotions that people create, enjoy and use; what cultural geographer Jon Anderson (2015) refers to as 'traces'. More formally, the anthropologist Clifford Geertz defined culture as '... an historically transmitted pattern of meaning embodied in symbols, a system of inherited conceptions expressed in symbolic forms by means of which men [sic] communicate, perpetuate and develop their knowledge about and attitudes towards life' (Geertz, 1973: 89). Culture, then, just like climate, is hard to see and harder to measure. As Tim Ingold says, 'We can never expect to encounter culture "on the ground"' (Ingold, 1994: 330), just as no-one has ever 'seen' climate. Instead, what we find are '... people whose lives take them on a journey through space and time in environments which seem to them to be full of significance, who use both words and material artefacts to get things done and to communicate with others, and who, in their talk, endlessly spin metaphors so as to weave labyrinthine and ever-expanding networks of symbolic equivalence' (Ingold, 1994: 330). It is therefore more accurate to say that people 'live culturally' rather than that they 'live in cultures'. Cultural geography, then, is the sub-discipline of geography which explores how cultural activities vary from place to place; indeed, it recognises how the very idea of 'place' is constituted through cultural practices. In *Weathered: Cultures of Climate* I am making the claim that the idea of climate exists at the intersection of culture, weather and place.

---

## The Structure of the Book

The book is organised into three sections. In Part 1, 'Knowledges of Climate', I explore the range of knowledges which humans have developed both historically and geographically about their climates. These cultural knowledges embrace different theories of why the physical properties of climates seem to change over time. In Part 2, 'The Powers of Climate', I illustrate how climates and cultures interact in specific places to shape patterns of life and how the idea of climate engages with different social practices and imaginative worlds. For example, climate enters into cultural accounts of blame and fear in many ways and finds representation in many cultural forms. In Part 3, 'The Futures of Climate', I consider the ways in which the futures of climate and humanity are inescapably bound together. On the one hand, different cultures construct a variety of trustworthy knowledges of how climates may change in the future. On the other hand, such knowledge of the future already changes the ways in which humans live in the present, and the idea of climate, I argue, continues to evolve.

The chapters in each of these respective sections are briefly summarised below.

## Part 1: Knowledges of Climate

Cultures in different historical eras have engaged with the idea of climate in a surprising variety of ways (**Chapter 2: 'Historicising Climate'**). Climates have been constructed from a wide range of imaginative and material evidence. These constructed climates have then been brought into public life to discipline personal, social and political behaviour in contrasting ways to diverse ends. The idea of climate has been bound up with, *inter alia*, imperial power, chauvinism, identity, nationhood, diet, colonialism, trade, health and morality. Precursors and parallels to contemporary thinking about climate can be found in earlier cultures' interpretations of their climate, while novelties and peculiarities can also be found which both challenge and disturb. These cultural histories of climate demonstrate that changes in the conceptual and rhetorical meanings of climate can and do exert a significant influence on public life.

What is known about the climates in which people dwell is always hard-earned, whether it be first-hand personal knowledge of the weather, second-hand knowledge of local climate that is held in cultural memory or scientific knowledge of changing climates acquired third-hand from trusted sources (**Chapter 3: 'Knowing Climate'**). What people know about their climate is also influenced by the cultures of meaning into which they are born and by the cultural practices of knowledge-making through which they become disciplined as citizens, practitioners, artisans or scholars. All knowledge of climate is cultural; it cannot exist separately from the cultures in which it is made or through which it is expressed. But which knowledge claims people deem to be trustworthy and which are held to carry public authority is not just a cultural question; it is also a deeply political one.

Human anxieties about a disorderly climate are long-standing. Climate is an idea which performs important functions in stabilising relationships between the experience of weather and cultural life. So when physical climates appear to change, the search for explanation becomes pressing (**Chapter 4 'Changing**

Climates'). Three categories of explanations for such changes can be discerned: the supernatural, the natural and the human. These explanations co-exist in complex ways within and across different cultures and there is an ebb and flow to their respective cultural authorities. But it is unusual for humans to think that climates change for either natural or supernatural reasons *alone*. Far more common, and indeed perhaps more necessary, is to believe that the performance of climate is tied to the behaviours of morally-accountable human actors. For much of the past, and in most places, climate and humans have been understood to move together, their agency and fate conjoined through the mediating roles of natural processes and supernatural beings.

## Part 2: The Powers of Climate

The idea of climate can fruitfully be understood as emerging from how people live materially and imaginatively with weather in particular places (**Chapter 5: 'Living with Climate'**). Climate becomes a rich ensemble of atmospheric processes, material technologies, memories, landscapes, dress codes, social practices, symbolic rituals, emotions and identities. Taken collectively these climatic behaviours may be thought of as 'weather-ways'. Patterns of weather are of course diverse around the world, and many of these patterns are now changing in significant ways. Yet how humans make sense of their weather and its changes cannot be separated from how history, the body and the imagination are expressed in specific places. Wider reflections taking place in the disciplines of geography, psychology and sociology around notions of place, identity, perception and social practice help to understand these sense-making cultural processes.

Climate has frequently been bound up with narratives of blame (**Chapter 6: 'Blaming Climate'**). There is a long history of elevating climate as the (primary) determinant of human physiology and psychology, just as there is one in which climate is offered as the primary determinant of physical landscapes, biological evolution and economic prowess. Wars, economic performance, street violence, political despots, famine, property prices, suicides, the age of menstruation – and many more phenomena – have all been 'explained' by climate. Culturally credible and persuasive accounts of blame and culpability fulfil an important human social and psychological need. Such collective sense-making of a complex and chaotic world enables social institutions to function and societies to be governed.

Ecological disorder and fear of the unknown future are enduring sources of human anxiety. In **Chapter 7: 'Fearing Climate'** I explore what happens to human emotions when either the experience of past climatic disorder or the claims of a future descent into climatic chaos feeds the imagination in powerful ways. Although fearful interpretations of climatic behaviour are common throughout human history, these fears are always mediated culturally. Climatic fears are bound up in wider narratives of apocalypse, risk society, emergencies and psychological (in)security. Climate chaos therefore exists in the imagination

as much as it can be discovered through scientific instrumentation and calculation. And however contemporary climatic fears have emerged, they will in the
end be dissipated, reconfigured or transformed as a function of cultural change.

Climate is an imaginatively fruitful idea and so it is inevitable that it will
be represented in different ways (**Chapter 8: 'Representing Climate'**). But representing the idea of climate is a challenge both for scientific practice and for
the arts. Climate and its changes are beyond mere mimetic representation,
whether by computer simulation or photography. No single timeless truth about
climate and what it means for people waits to be revealed through science or
through art. Despite frequent exhortations to the contrary, climate science cannot 'demand' any particular course of action of people. Neither can art. It is not
didactic, it cannot instruct people in sustainable, just or alternative living. Since
there can be no unmediated access to climate, all representations of climate are
in the end political acts; that is, they are engaged in constructing different and
selective climatic realities: material, ideological, imaginative, normative. But
good climate art will engage human faculties to provoke reflection on the profound questions prompted by foretold changes in climate: the good life to be
admired, the future to be aspired to and the responsibilities they have to others,
both human and non-human.

## Part 3: The Futures of Climate

There is a long cultural history of claims-making about the future, scientific
forecasting being only the latest in the tradition of prophetic knowledge
(**Chapter 9: 'Predicting Climate'**). As with all predictive knowledge, climatic
predictions are culturally conditioned and how such predictions are used varies from culture to culture. Predictions of future climate, whilst often relating
in some way to scientific knowledge claims, are always mediated by a wider
variety of cultural norms, computational artefacts and communicative practices. Numerical computer modelling of a climate is but one means of
predicting its future course, but one that has taken centre stage in the contemporary world. Yet model predictions, and those who communicate them –
standing in a long line of prophetic voices – face multiple challenges: how to
be credible, how to carry authority, how to be useful. Navigating between
these demands is not easy. Predictions will nearly always turn out to be wrong;
the future has its own unfolding logic of action beyond human reach. But
cautionary tales about the future are absorbed, pondered and can usefully be
learned from.

People have long sought ways to make their climate more agreeable,
whether by migrating in search of more favourable ones or by adapting to
climate's vicissitudes. The idea of deliberately redesigning climate is more
recent and can be traced to modernist European projects of improvement and
colonisation (**Chapter 10: 'Redesigning Climate'**). Technological entrepreneurs
of the nineteenth and twentieth centuries aspired to 'control the weather'.
More recently the idea of global climate engineering has emerged. With global

temperature as the dominant metric for revealing climate-change, it has become the control variable upon which human designs might be wrought. Climate engineers today are proposing and researching technologies that would 'manage' global temperature, much as the householder seeks to manage the comfort of her home or car by turning the thermostat. But climate engineering raises profound philosophical and political questions. What would it be like for people to live in an enhanced or restored climate, knowing that it was made and maintained by human hands? And, politically, who gets to decide and govern the climates thus designed: those who have most to gain, those who have most to lose or those who simply have the power and means to enact the technology?

Climate has long been a political object around which different modes of governing, regulating and ordering society vie for recognition and endorsement (**Chapter 11: 'Governing the Climate'**). But when climate is to be governed, what precisely is it that is being steered? The cyclones, heatwaves, ice-storms and downpours that begin to constitute the physical and imaginative contours of climate are not directly subject to human laws, policies or technologies. Governing climate therefore always becomes a project about governing and controlling things other than the weather: physical environments, social practices, material technologies, investment flows. Governing local or colonial climates has usually been an exercise in governing local communities or colonial societies. And the rise of *global* climate governance in recent decades has further extended the range of practices, technologies and institutions which can come under the reach of climate governance. Governing global climate becomes an exercise in governing global society, but where the power to do so exists in no central or identifiable locus.

In the last chapter of the book (**Chapter 12: 'Reading Future Climates'**), I speculate about the future of climate as an idea. I offer three possibilities. First, is the idea of climate re-secured within desirable and 'safe' limits. This ambition seeks to shore-up the historical function of climate by re-establishing a degree of orderliness in the world. A second possibility embraces a future of improvised climates rather than of re-secured or stabilised ones. Improvisation recognises that physical climates will always escape human management. A third possibility is more radical and calls into question the imaginative function of climate upon which this book is premised. It suggests that in the new epoch of the Anthropocene people may have to learn to live without the idea of climate, at least without climate as an idea that brings order and stability to relationships between weather and culture. The 'new normal' of climate is simply that there can be no normal. And this is unsettling.

## Chapter Summary

The idea of climate is as old as the human imagination and, as an imaginative way of bringing order and stability to human life, the idea remains pervasive across today's diverse cultures. Climate *may* be defined according to the aggregated statistics of weather in places or as a scientific description of an interacting physical system. But climate may also be apprehended more intuitively, as a tacit idea held in the mind or in cultural memory of what the weather of a place 'should be' at a certain time of year. However it is defined, formally or tacitly, it is people's sense of climate that establishes certain expectations about the atmosphere's performance. The weather experienced often fails to meet people's expectations, but the fact that they *have* expectations is due to the idea of climate. As a normalising idea climate therefore offers people a certain sense of security; it allows them to 'put weather in its place'. Climate introduces a sense of stability and normality into what otherwise would be too chaotic and disturbing an experience of unruly and unpredictable weather. Climate cultivates the possibility of a stable psychological life and of purposeful action in the world.

Put simply, the idea of climate allows people to live culturally with their weather. Climate is weather which has been cultured, interpreted and acted on by the imagination, through story-telling and using material technologies. And as I now proceed to explain in the rest of the book, both people and their cultures are in continual processes of being weathered.

## Further Reading

Behringer, W. (2010) *A Cultural History of Climate*. Cambridge: Polity Press.

Bristow, T. and Ford, T.H. (eds) (2016) *A Cultural History of Climate Change*. Abingdon: Routledge.

Leduc, T.B. (2010) *Climate, Culture, Change: Inuit and Western Dialogues with a Warming North*. Ottawa: University of Ottawa Press.

# Part 1
## Knowledges of Climate

# 2

# Historicising Climate

## Introduction

Ideas about climate are always situated in a time and a place. Classical Greeks understood climate – *klima* – differently to earlier Jewish conceptions of the weather. Early Enlightenment philosophers or today's scientists understand climate differently to either of these earlier cultures, as do many non-western worldviews continue to conceive of climate today (see **Box 1.1**). Climate means different things to different people in different eras and in different places. Given the importance attached in today's world to the phenomenon and discourse of climate-change[1], it is important to develop accounts of the changing historical meanings of climate. In other words, it is necessary to historicise climate.

To historicise an idea is to interpret it as the product of historical development, to recognise how and why its meanings have changed over time. Contemporary climate-change discourse is focused on the changing *physical* attributes of climate and what this might mean for social and ecological futures. Yet it is equally important to recognise that the *imaginative* dimensions of the idea of climate – what climate signifies to people; in other words its cultural meanings – also change and often in substantial ways. As anthropologist Michael Dove observes, 'The engagement by the climate science community with history to date has been with the history of [physical] climate, not with the history of ideas about climate' (Dove, 2015: 28). This chapter is concerned with this latter objective.

Yet my interest in historicising climate is not to seek and acquire some deep, essential meaning of the word itself. Political geographer John Agnew has warned that a focus on the intellectual history of words for its own sake may be misguided. Rather, he says, '... tracing the meanings of ... words historically is valuable insofar as it can alert us to how some, often highly selective, meanings have changed. It is insufficient [by itself] to tell us how relevant the word now is to what we are currently interested in defining'

---

[1] See Note in Preface on p.xii for an explanation of how I use this expression throughout the book.

(Agnew, 2014: 312). What matters is how the idea of climate is understood and used relative to a certain time and place. There is no single true and eternal definition of climate to be discovered or defended. Meanings of 'climate' have indeed changed historically; and they will change again. They have also differed culturally. Being alert to the effects that the residues of such different meanings of climate continue to exert in today's world is a state of mind that this book seeks to cultivate.

## The Many Uses of Climate

As well as being an old idea, climate is also a versatile one. While one can imagine an unbroken sequence, long pre-dating humans, of moment-by-moment *weather* unfolding in the Earth's atmosphere, *climate* is an idea invented in the human mind. Although Classical Greek philosophers were perhaps the first to leave an articulate written account of climate, they would not have been the first people to seek to make sense of the incessant flow of atmospheric phenomena. The rhythms of the sky have always been companions of human thought and triggers of ritual. From the frigid north to the torrid tropics such rhythms have induced wonder and fear, whilst also offering comfort and assurance. Alongside the experience of intense yet predictable diurnal and seasonal weather cycles, has sat the unreliable performance of the atmosphere from year to year (no two years are the same) and from generation to generation (the weather of old age seems unlike the weather of youth). While a drought is to be feared, a mere dry season is not. A winter is not an ice age; neither are all storms hurricanes. It is little wonder that human anxieties, hopes and the search for explanation, and hence many spiritual longings and theologies, have been bound up with the skies.

Religions have found many ways to make sense of these cruel fates delivered from above, acknowledging dependence on powers beyond human control and giving thanks for mercies and blessings received. Climate and religion therefore have a long history of interdependence (Botero et al., 2014; Hulme, 2016a). For example, the Jewish scriptures reveal that the Israelites of the first millennium BCE had expectations about the regularity and appropriateness of the weather – 'Like snow in summer or rain in harvest, honor is not fitting for a fool' (Proverbs 26:1) – even if they did not articulate them linguistically using the specific idea of climate. And their theory of weather was integral to their sense of moral accountability and divine providence. Thus the Book of Job reports, 'At his direction [the clouds] swirl around over the face of the whole earth to do whatever [God] commands them. He brings the clouds to punish people, or to water his earth and show his love' (Job 37:12–13).

More generally through history, the idea of climate has worked both as an index and as an agent. Climate has been used both to describe (e.g. patterns of weather) and to explain (e.g. ecological, social or human outcomes). As *index*, climate describes the accumulated rhythmic patterns of weather in places.

Thus the original Greek notion of *klima* was a description of latitudinally varying solar inclinations which (roughly) correlated with air temperature. And writing in the early seventeenth century William Shakespeare could describe Sicily's climate in *The Winter's Tale* as 'delicate' and 'sweet'. In the modern era too, numerous climatic indices – more quantitative than Shakespeare's adjectives – are popularised as short-hand descriptions of the time/space averaged state of the atmosphere. For example, the El Niño/Southern Oscillation (ENSO) and global-mean temperature are both descriptive indices of larger-scale physical systems. As *agent*, on the other hand, climate is used to explain a bewildering range of physical, social and human outcomes: landscapes, species extinctions, wars, suicides and many more phenomena (explored further in **Chapter 6**). This usage of climate was also familiar in Classical Greek and Roman thought, where the idea of climate offered an explanatory framework for human cultural diversity and acted as a moral guide for geographical exploration.

While this dual function of climate has therefore recurred throughout human cultural history, 'Climate has more often been defined as what it *does* rather than what it *is*' (Fleming and Janković, 2011: 2, emphasis in the original). According to these historians climate has been understood more commonly as an agent than it has as a description of aggregated weather. Climate's agency, they go on to suggest, has extended historically 'as a force ... informing social habits, economic welfare, health, diet, and even the total "energy of nations"' (2011: 2). This 'forceful' role for climate can be seen today, for example, through the UN's IPCC devoting an entire Working Group to evaluating the agential efficacy of climate change (impacts) and how humans can dissipate this force (adaptation). The distinction between climate as index and climate as agent is crucial for understanding the cultural role played by the idea of climate in today's world. The distinction is also important for appreciating both the material and imaginative manifestations of climate.

\* \* \*

The rest of this chapter is organised around four different categories of historical climatic discourse: climates as civilised, as national, as commercial and as problematic. Together, these illustrate some of the many ways in which climate has been understood historically within different cultures.

## Civilised Climates

In his classic work on the history of nature and culture in western thought, Clarence Glacken suggests that the complex relationship between climate, human behaviour and culture has been one of the enduring questions to engage the mind (Glacken, 1967). One of the preoccupations of these reflections has been to establish which climates are most conducive to virtuous human character, to prosperous economies and, ultimately, to the highest forms of civilised life. The answer has usually been the climate coincident with the one in which the questioner dwells! For the Greeks, climates to the north were too cold and those in the south too hot to allow significant human cultural development. But like the children's story 'Goldilocks and the Three Bears', it was the climates of the intermediate Mediterranean zone which were 'just right' for the blossoming of cultural creativity. Thus the Greek philosopher Hippocrates writing in the fifth century BCE could claim,

> I hold that Asia [Minor] differs very widely from Europe in the nature of all its inhabitants and all of its vegetation. For everything in Asia grows to far greater beauty and size; the one region is less wild than the other, the character of the inhabitants is milder and more gentle. *The cause of this is the temperate climate*, because it lies towards the east midway between the risings of the sun, and further away than is Europe from the cold. (Hippocrates, 1923: xii; emphasis added)

The 'hospitable' climate of the Aegean Sea became a way of simultaneously explaining and justifying Greek hegemony in the Classical world. Similar reasoning can be found in Arab literature from the fourteenth century CE, for example, the Tunisian geographer Ibn Khaldun, and in the writings of late Renaissance and early modern European thinkers. Immanuel Kant lecturing in Prussia in the second half of the eighteenth century therefore echoed Hippocrates when he claimed that,

> The inhabitant of the temperate parts of the world, above all the central part, has a more beautiful body, works harder, is more jocular, more controlled in his passions, more intelligent than any other race of people in the world. That is why at all points in time these peoples have educated the others and controlled them with weapons. (Kant, *Physical Geography*, cited in Livingstone, 2004: 77)

The only difference for Kant of course was that the hospitable temperate climate of the Greeks had shifted 2,000 kilometres northwest from the Aegean Sea to the Baltic coast.

Perhaps the most developed articulation of this line of reasoning came from the American geographer Ellsworth Huntington in the early decades of the twentieth century. Drawing upon empirically based studies examining the productivity of factory-workers under different climatic conditions, Huntington reaffirmed what Hippocrates, Ibn Khaldun and Kant might have argued, namely that 'the denizens of the torrid zone are slow and backward, and we almost universally agree that this is connected with the damp, steady heat' (Huntington, 1915: 2). The locus of civilising climate had now shifted to the eastern coast of North America, where Huntington lived, and to the bracing climate of northwest Europe. It was a short step from this climatic theory of racial hierarchy to arrive at a climatic basis for the shifting fortunes of the ancient civilisations of the Near East, Mediterranean and central America. In his 1915 book *Civilisation and Climate*, Huntington devoted two chapters to exploring the idea that the fate of civilisations is intimately related to physical changes in climate. He concluded that, 'no nation has risen to the highest grade of civilisation except in regions where the climatic stimulus is great … *a favourable climate is an essential condition of high civilisation*' (Huntington, 1915: 270; emphasis added). If such views of climate strike us today as naïve or even dangerous, they also point to the importance of historicising the idea of climate. Huntington's intellectual legacy endured through the decades of the twentieth century, even if not always expressed quite as stridently, and it remains alive in the present day (Powell, 2015); see also **Chapter 6**.

## National Climates

The idea of climate has also served in various ways to forge or defend ideas of national character and identity, even to contribute to emergent claims of nationhood. Giving a credible account of how and why 'the Other' is different is central in all forms of human identity formation. Thus one can listen to Herodotus in the fifth century BCE making sense of – at least for himself – why Egyptians are not Greeks, 'in keeping with the *idiosyncratic climate* which prevails there and the fact that their river behaves differently from any other river, almost all Egyptian customs and practices are the opposite of those of everywhere else' (Herodotus, 1998: 35; emphasis added). Or again, deferring to Shakespeare, one can see how descriptions of climates can be bound up in (at least imagined) notions of national character and identity (see **Box 2.1**). If not the superiority of nations, then certainly differences among peoples have been ascertained by the discerning reader of national climates.

---

### Box 2.1: Shakespeare's Climate

In late sixteenth-century England, climate was a signifier of ominous events, of human moral failure and of judgements of the divine will. In the lives of Shakespeare's contemporaries, climate carried emotional, spiritual and discursive properties more than it revealed the working of natural physical systems. As well as revealing the portentous and idiosyncratic events of daily life, the ordering and differencing effects of the idea of climate also found resonance in an age of geographical exploration and exploitation. The idea of climate helped to order, and therefore to make sense of, differences between regions and countries. The forays of sixteenth-century European adventurers into extra-European longitudes were beginning to challenge the old Greek idea that fixed and identical climates exist at different latitudes.

*(Continued)*

*(Continued)*

New climatic schemes were needed to explain the novel and exotic climates of the Caribbean, the eastern seaboard of the Americas and the monsoonal climates of India. These travellers brought back tales of climates unknown to the island dwellers of northwest Europe and a new geography and vocabulary of climatic differentiation was needed.

Shakespeare would have been fully aware of these new encounters with distant and novel climates and the commercial prospects that they offered. Foreign and exotic climates could be both feared and valorised by Elizabethans. The Earl of Essex and his 1599 military expedition feared the Irish climate would 'consume our armies, and if they live, yet famine and nakedness makes them lose both heart and strength' (Shapiro, 2009: 90), while Shakespeare could imagine the enticements of Mediterranean climates he must have heard much of, referring to Sicilia (the island of Sicily) in *The Winter's Tale* where 'The climate's delicate, the air most sweet'. Yet if Sicily's climate was delicate, Shakespeare could project a very different image of an English climate that was 'raw and dull'. In *Henry V*, Shakespeare adopted the ancient trope of using climate to reveal national character. As the French contemplate facing up the English army, he has the Constable of France speak in bewilderment of the relationship between England's climate – 'foggy, raw and dull' – and her people's character: 'Dieu de batailles! where have they this mettle? Is not their climate foggy, raw and dull, On whom, as in despite, the sun looks pale, Killing their fruit with frowns?'[2]

(Sourced from Hulme, 2016b)

The possibilities of using the idea of climate in projects of nation-building were given further impetus in eighteenth-century Europe as weather was, for the first time, measured systematically. The physical properties of the atmosphere were increasingly classified through meteorological instruments. These properties were recorded initially through individuals' observations of the weather in private diaries and then later through systematic and bureaucratic networks of observing stations and tabulated registers. These standardised methods of observing the atmosphere opened up new ways of describing climate. Order was imposed on seemingly chaotic weather, first, by quantifying it locally at individual places and, subsequently, by constructing statistically aggregated climates from geographically dispersed sites. Climate, for the first time, became 'domesticated' (Rayner, 2003) and then, through international scientific cooperation in the nineteenth century, mapped across national boundaries. Following Alexander von Humboldt's cartographic innovation of isothermes in the early nineteenth century, regional and eventually global climatological maps of various sorts could be produced connecting together disparate places which shared similar temperature or rainfall.

---

[2] Shakespeare, *Henry V*, III, v

These innovations facilitated new quantitative exercises of delineating one type of climate from another and of comparing numerically the climates of nations and places. Such projects suited the forging of political identities and the establishment of European colonies. These numerical climatic comparisons served various goals. For example, they helped construct a sense of solidarity amongst inhabitants of specific places and political jurisdictions through the shared experience of a similar climate. Historian Jan Golinski has shown how during Britain's (eighteenth) 'century of enlightenment' certain ways of thinking and talking about the weather and climate became embedded in British national culture (Golinski, 2007; see **Figure 2.1**). Standardised measurements of weather through instrumentation, combined with emotional and cultural idioms, helped to develop a shared sense of national British climate as one that was temperate but punctuated by invigorating diurnal variations: 'The British came to see their national climate in a much more favourable light, appropriating for themselves the temperate ideal that the ancients had assigned to the Mediterranean' (Golinski, 2007: 56).

**Figure 2.1**  'Delicious Weather'. One of James Gillray's 1808 series of cartoons showing typical British characters in various weather conditions. Here the national character finds its ideal weather conditions, associated with agricultural fertility, good health and flourishing wildlife (Source: Golinski, 2007).

Numerical climatic comparisons also enabled clearer distinctions to be drawn between the climates of nations. Whereas Shakespeare had placed in the mouth of a Frenchman the notion of England's climate as 'foggy, raw and dull', by the late eighteenth century it was possible to say exactly how much foggier and duller London's climate was compared to that of Paris. Nation-building in an imperial era also benefited from these new conceptions of climate as measurable and precise. By 1800 it was possible to contrast British climate with the more exotic and extreme climates encountered by western-moving settlers in North America. Historian Golinski again: 'Whether perceived as too hot or too cold, it was the American climate's differences from the British climate that aroused anxieties about sickness' (2007: 195).

Of course eighteenth- and nineteenth-century European settlers in America had difficulty coming to terms with their new climates. They did not know what to expect from the continental climates being newly encountered as they moved westward across the interior. They were as likely to be guided by analogues imported from more familiar climates, such as Europe and Russia, as they were to develop climatic knowledge based on their own long-term systematic observations, which were of course initially lacking. They had neither personal or cultural memory nor instrumental measurements to develop their sense of American climate. What *was* clear was that the Classical conception of latitudinally determined *klimata* was grossly deficient as a model of climatic behaviour: the climate of the American West was nothing like the American East, let alone the climate of Europe. The struggle to invent America's climate became caught up in arguments about westward territorial expansion, about the spread of slavery and cotton, and about the effect of human land clearance on local climates (Culver, 2014). Climates then, whether real or perceived, have not only shaped national identities but can themselves be influenced by the cultural imagination.

## Commercial Climates

Climate can exert its imaginative influence not only over cultural identities, but also as a way of promoting and disciplining economic growth. A third set of discourses which helps illustrate the changing conceptions and uses to which the idea of climate has been put therefore concerns the connection between climate and economic activity: resource extraction, commerce, trade and wealth production. These concerns became increasingly important to the European colonising powers of the sixteenth to twentieth centuries and introduced new ways of evaluating climate. Beyond its relationship with place, character and civilisation, the idea of climate was conceived and utilised to further economic trade and to justify new investments in cultivation and resource extraction. Climate became 'productive' of biological or geophysical qualities which had economic value. Alongside power, capital and labour, climate became a factor of production.

Vladimir Janković (2010) has shown how this thinking manifested itself in one specific case in the first half of the eighteenth century. The Swiss trader and entrepreneur Jean Pierre Purry singled out the climate of South Carolina as 'optimal' for settlement, resource extraction and mercantile trade. Purry was able to construct discursively a 'model climate' for the region in order to advance his own personal political and economic project. On the back of loans raised in London and Paris, he founded the township of Purrysburgh in the 'ideal' climate from which to launch his new commercial plantation projects. While this Carolian climatic ideal lived largely in Purry's imagination, it nevertheless carried coercive power when projected through the rhetoric of a persuasive businessman in front of European money-lenders.

Although idiosyncratic, and ultimately a failure, Purry's thinking was to be replicated many times over in the centuries to follow. European colonisers sought to establish numerous agricultural projects and commercial plantations around the world, guided by notions of fruitful and productive climates. The quantitative descriptions of transnational climates in the nineteenth century mentioned above were also to prove economically valuable in these imperial projects of resource extraction and trade. Comparative climatic analysis of different regions could be undertaken, now relying on numerical data rather than on impressions or hearsay. How different was the climate of Cape Town from that of Amsterdam? Where best in the tropics and sub-tropics could wheat, cocoa or rubber be grown? Longitudinal studies of climate through time also became possible for the first time. These provided a formal alternative to the more limited reach of human memory and allowed the temporal reliability of newly colonised climates to be established.

Plantation forests were one example of the beneficiaries of this new climatic enterprise. Forests had great economic value in new European colonies and the exciting economic and ecological potential of newly 'created' and quantified climates could be realised. For example, climatic data were crucial for the South African Forest Service of the late nineteenth and early twentieth centuries. Earlier efforts to import *Eucalyptus* and *Acacia* seeds from Australia for huge new plantations on the largely treeless southern African plains and hills had mostly failed until the English forester David Hutchins arrived there in 1881. Hutchins was fully appraised of the new climatic statistics which were being generated across the British Empire, not least in Australia and South Africa. He developed a simple mantra which was to revolutionise forestry practice in the country: 'fit the tree to the climate'. Trees from anywhere in the world could be raised in South Africa so long as careful attention was paid to the relevant climatic conditions. As Hutchins wrote to the agriculture minister in 1897, 'In South Africa with its variety of trees and climates, meteorology and the climate requirements of each tree are the most important study for foresters' (quoted in B.M. Bennett, 2011: 273).

This association between climate, commerce and economic opportunity remains visible today. This is not simply in the sense that climatic information has commercial value, whether in guiding investments in agriculture, infrastructure or in a wide variety of other economic projects. But the ideas also persists, traceable back through earlier eras, that some climates lend themselves more 'naturally' to economic prosperity than do others. There are some climates in which economies seem to prosper; there are others that seem innately problematic.

## Problem Climates

The American climatologist Glenn Trewartha published a book in 1961 called *The Earth's Problem Climates*. 'Problem climates' for Trewartha were those which did not conform either in location or in magnitude to the idealised world climatic pattern which might be expected from the laws of physics applied to a hypothetical continent. Thus the dry west coast of South America, the rainfall regimes of the Guinea Coast in Africa and its hinterlands, and the dry region of the equatorial Pacific were all 'problem climates' for Trewartha.

This rather idiosyncratic identification of problem climates saw them purely in abstract scientific terms. Far more common in the history of climatic thought has been the problematisation of climates according to what they appear to do or not do for people. A common distinction to make about the role of climate in human affairs has been between seeing climate as a resource, a constraint or a hazard. Problem climates become those which are deemed hazardous for human health or endeavour, or else those which impose some assumed constraint upon desirable economic or social activity. It is this latter notion of problem climates that seems implied in those arguments which claim that economic productivity is innately limited by tropical climates. From Adam Smith in the late eighteenth century onwards, and also before, many have claimed that the noticeable geographical variation in 'the wealth of nations' bears some approximation to the distribution of average temperature. Take this recent example from a leading American development economist: 'The very poorest regions in the world are those saddled with both handicaps: distance from sea trade and a tropical or desert ecology' (Sachs et al., 2001: 71), where for Sachs 'tropical or desert ecology' is short-hand for a problematic climate. Or more bluntly Burke et al. (2015: 235) claim that '... [economic] productivity peaks at an annual average temperature of 13°C and declines strongly at higher temperatures'. Here perhaps are echoes of Huntington's denizens of the torrid zone who are 'slow and backward'.

But it was concerns about human moral and physical health that most frequently contributed to the discourse of problem climates. Such judgements were, of course, all heavily influenced by culture. For Europeans of the nineteenth century, and earlier, tropical climates were frequently deemed to be dangerous. The encounter between white settlers and the unknown climates of south Asia, Africa and South America invoked fears and anxieties, whether due to the potential for degeneracy, depravity or debility. These newly encountered climates took on the roles ascribed to them by the prevailing and dominant culture, as has so often been the case throughout history. The moral classification of tropical climates as dangerous and threatening was tightly bound up in the discourse around acclimatisation: could white Europeans settle, survive and rule in 'hostile' climates? The association of (tropical) climate with fear, danger and anxiety was as much a function of the imperial ideology of the day as it was a function of objective physical or medical diagnosis. Opinion became polarised in the later Victorian period about whether or not the unknown and forbidding climates of the tropics were to be feared and were thus in need of 'conquering'. For example, in 1898 the Anglo-French medic Dr Luigi Westenra Sambon addressed the Royal Geographical Society in London and sought to persuade his late-Victorian audience that the pathologising effects of tropical climate had been exaggerated. Sambon was an apologist for colonial settlement and argued that European settlers were quite capable of adapting to various tropical climates by paying careful attention to the discoveries of medical science (Livingstone, 1999).

Tropical climates *were* eventually 'conquered' in the European mind, in a metaphorical as well as a literal sense. For example, it was widely regarded that sustainable colonisation of India by Europeans required periodic escape by the settlers to the cooler climates of the Indian hills, stimulating the construction of hill stations as white enclaves (see **Box 5.2** in **Chapter 5**). Later improvements in tropical medicine and air-conditioning technologies removed some of the direct physical fears tropical climates presented to non-indigenous populations. As imperial projects disintegrated during the twentieth century, the psychological hold on the European mind of the pathology of tropical climates weakened (see **Box 2.2**). Yet at various stages in climate's cultural history, concerns about 'problem' climates whether for commercial, moral or health reasons have given rise to a desire for climatic improvement or enhancement. This is an idea I examine in more detail in **Chapter 10**.

---

## Box 2.2: Reinventing Caribbean Climates

For early modern Europeans tropical climates were often understood as morally degrading, disease ridden and deadly. This was true of many eighteenth-century conceptions of the climates of Caribbean islands. Many writings from the time, for example, James Johnson's 1818 book *The Influence of Tropical Climates on European Constitutions*, emphasised the importance of locality for determining the (un)healthiness of tropical climates. A combination of a meridian sun, marshy rotten soil and heavy rains made it impossible for Europeans to think of Caribbean island climates as anything but unhealthy. But changes in commerce, medical knowledge and, especially, the rise of international tourism had by the twentieth century altered Caribbean climates in the eyes of most Europeans into a desirable commodity. Caribbean island climates, with sunshine, warmth and predictable dry seasons, began to be marketed for foreign consumption. For example, a 1905 hotel brochure from Barbados claimed it was located in 'the most ideal winter resort of the tropics for tourists, invalids and those seeking a genial climate' (quoted in Carey, 2011: 139). As Carey remarks in his analysis of these changing conceptions of the Caribbean, 'The warm tropical climate, far from a deadly

*(Continued)*

*(Continued)*

deterrent, was now the Caribbean's best economic asset' (2011: 136). Caribbean climate 'changed' dramatically during these two centuries, from being deadly and disease-ridden to become one of the world's most idyllic tourist havens for sun-starved tourists.

(Sourced from Carey, 2011)

## Chapter Summary

In this chapter I have shown how cultures in different historical eras engaged with the idea of climate in a surprising variety of ways. Climates have been constructed from imaginative and material evidence – both indirect and sensory – and these constructed climates have then been brought into public life to discipline personal, social and political behaviour in contrasting ways to diverse ends. Climate has functioned historically both as index (of weather in places) and as agent (of physical change and social outcome). The idea of climate has been bound up with, *inter alia*, imperial power, chauvinism, identity, nationhood, diet, colonialism, trade, health and morality. These examples show how, in different historical settings, the idea of climate has brought order and stability to the agitated relationship between weather and human cultures. Precursors and parallels to contemporary thinking about climate can be found in earlier cultures' interpretations of their climate, while novelties and peculiarities can also be found which both challenge and disturb. These cultural histories of climate demonstrate that changes in the conceptual and rhetorical meanings of climate exert significant influences on public life. This shaping of culture by the imaginative power of the changing idea of climate is perhaps as great an influence on human life as might be accomplished alone by changes in the physical properties of climate. The idea of climate retains tremendous power and utility in today's world and in the next chapter I examine some of the different forms of climatic knowledge that are in public circulation.

## Further Reading

Barnes, J. and Dove, M.R. (eds) (2015) *Climate Cultures: Anthropological Perspectives on Climate Change*. New Haven, CT: Yale University Press.
Golinski, J. (2007) *British Weather and the Climate of Enlightenment*. Chicago, IL: Chicago University Press.
Janković, V. (2000) *Reading the Skies: A Cultural History of English Weather, 1650-1820*. Manchester: Manchester University Press.

# 3

# Knowing Climate

## Introduction

It is not easy to make knowledge about the world. Even harder is it to ensure that knowledge that *is* made remains stable over time. Equally, one can never assume that knowledge about some phenomenon which appears sure and secure in one place will carry the same cultural authority once it is articulated in a different setting. Scientific knowledge is held in high regard in most western European societies, and elsewhere too. But religious or indigenous knowledge might carry equal or greater authority in some national cultures or in some ethnic sub-cultures. And what might be called personal, or tacit, knowledge is important everywhere. To inquire into the nature of knowledge – the discipline of epistemology – is to ask what does it mean when someone claims to know something? How and why does one person's knowledge differ from another? Answering such questions is not a trivial task (Burke, 2008). To study knowledge is also to ask how people use their reason, imagination and senses to acquire knowledge, or else to ask why people rely on the testimony of others. To be sceptical of assumed knowledge is to claim that it is possible that people do not know as much as they think they do.

These characteristics and questions of knowledge in general, apply equally to knowledge about climate and its changes. What is known about the climates in which people dwell is always hard-earned, whether first-hand personal knowledge of the weather, second-hand knowledge of local climate that is held in cultural memory or scientific knowledge of changing climates acquired third-hand from trusted sources. What people know about their climate is also influenced by the cultures of meaning into which they are born (what an Inuit seal-hunter knows about climate will be different to what a Chinese city-dweller knows) and by the cultures of knowledge practice through which they become disciplined (a climatologist knows climate differently to an anthropologist). All knowledge of climate should therefore be regarded as cultural: it cannot exist separately from the cultures in which it is made or through which it is expressed.

All knowledge of climate is also political. Or, to be more precise, the forms of climatic knowledge which become authoritative and trusted in a given society are a result of political processes. Different political actors, shaped by different cosmologies, ideologies and values, will hold different views as to what counts as valid evidence upon which climatic knowledge claims are based. They will also hold different views of the trustworthiness of the various institutions – whether government bodies, non-governmental organisations (NGOs), universities, media platforms or religious institutions – which bring forward knowledge claims about climate and its changes. It should not be assumed that scientific knowledge of climate carries universal cultural authority. As many scholars have pointed out (e.g. Plumwood, 1993; Wilson, 2012), religious globalisms, subaltern cosmologies and other forms of subjectivities are increasingly challenging knowledge that is based on secular rationality. This challenge also is as true for knowledge about climate as it is for other objects of knowledge. Indeed, some have argued that the hegemonic conditions under which scientific knowledge has come to dominate debates about climate-change need challenging. Jennifer Rice and geography colleagues have argued that

> Climate politics urgently needs to be repoliticised to include more democratic debate and argument, based on a wider discussion of values, norms and experiences. This requires among other things a *discussion of the politics of knowledge* underpinning our current political condition. (Rice et al., 2015: 254, emphasis added)

\* \* \*

In this chapter I illustrate some of the different ways that people in different places make useful and trusted knowledge of the climates in which they live. There are different ways of knowing climate – through personal encounter, cultural myths, scientific practice, artistic expression – which rely on different evidential standards: personal diaries, oral histories, statistical analyses, controlled experiments, computer modelling and so on. This chapter also explains why it is that different knowledges of climate carry quite different authority in different cultural settings. There is no single and universal way in which *all* people come to know their climate, whether the climate of their locality or the climate of the planet. For example, the very idea that there might be such a thing to be studied as 'global climate' and to be the object of political argument is an invention of modernity and the scientific mind. To explore these questions of knowing climate, the rest of this chapter is structured around four different but overlapping categories of climate knowledge: personal, indigenous, scientific and consensual.

## Personal Climate Knowledge

How do people come to know the climate of the place in which they live? First and foremost it is through their direct experiences of the atmospheric weather they encounter day by day. People live and work in places and each place has its own distinctive patterns of weather which its dwellers experience. Climatic knowledge for many people is not abstract or formal – as in statistical representation – but is personal and impressionistic. The climatic knowledge that many people hold is bound up with places, bodies and with social practices such as farming, fishing, gardening and recreation. For someone to 'weather' the climate is for them to grow into a place, to be emotionally and materially connected to it, to dwell in a place fully (see **Chapter 5**). Some people feel impelled to inscribe this knowledge in private weather diaries, which by definition brings the personal into understandings of climate (Adamson, 2015). Private emotions, triggered memories and sensual experiences of weather are blended into shorter or longer narrative or quasi-numerical forms which offer forms of personalised climate knowledge.

As the world becomes increasingly mobile and mediatised, people may increasingly encounter *multiple* climates – if only vicariously through TV, film and the internet. Nevertheless, place attachment remains very important for the human psyche and the experience of weather in locality remains central to most people's personal knowledge of climate. In traditional or non-urban cultures this knowledge of place-based climates might still revolve around livelihood extractions from land or sea or from the mundane experience of the seasonal cycle. Literary scholar Nick Groom explores this sentiment in his book about traditional English rural culture. He argues that, for some, experiencing the seasonality of weather provides reassurance, 'its rituals and rhythms can still help us to understand the natural world in more "natural" ways' (Groom, 2013: 29). But urban-dwellers, too, may find their personal knowledge of climate materialised in specific places. Through domestic gardens or public parks, for example, people witness the annual cycle of climate manifest in plant growth, flowering and decay. In such compact city spaces the rhythms of climate are given manifest form in the colours, shapes and mosaics of vegetation and the life forms it supports. Christine Corton's cultural biography of London's fog (Corton, 2015) offers insight into how knowledge of one iconic aspect of one city's climate is deeply attached to material and imaginative expressions of place.

Personal knowledge of climate is not only bound up with places, bodies and practices. It is also bound up with memory. Over time, remembered experiences of weather build expectations about what is usual or unusual for the time of year in a particular place. Through memories of past weather, people build a sense of the stability or volatility of their climate, a sense of how 'appropriate' for *this* place is the weather experienced today. Memory is integral to the mental construction of the idea of climate, to what each person deems to be normal or exceptional weather (see **Box 3.1**). But weather memories are not only individual and personal; they are shaped powerfully by collective and cultural processes. As explained by one weather historian,

> Extreme [weather] events that result in trauma, flood and the epigraphic records of such events, can also become the focus of community memorial and mourning. These different forms of recording, remembering and memorialising the past represent media through which climate is constructed in cultural memory across generations. (Endfield, 2014: 306–7)

## Box 3.1: Climate and Memory

For most people the idea of climate becomes reified through a rather unstructured assemblage of remembered weather. For example, my personal knowledge of British climate, and therefore my own expectation of how weather *should* behave, accumulates through my life story. Memories of earlier encounters with weather intrude into my present experience and interpretation of the skies. For example, I vividly remember being battered and bruised as a six-year-old by a violent April hailstorm on the cliffs of Devon; I remember the sun-drenched summer of 1976 through the heroics of the visiting West Indian cricket team. And imagined memories of past English weather also reach me vicariously through re-told stories of my father's and grandfather's lives: the sun-lit and carefree summers of farm-life in the 1930s in north Wales, the deep snows of 1947 in urban Leeds during post-war austerity. This is weather captured not as abstract statistics, but as indexed personal memories and through powerful evocations of fear and desire. My knowledge of British climate therefore becomes inescapably entangled with a multi-sensory account of my own – and others' – past in which imagination, emotion, place, culture and history engages with the physicality of heat, cold, wind, sun and rain.

Personal knowledge of climate is therefore profoundly embedded in memories of past weather: my own memories, those of my ancestors, those of my culture. Claims such as 'The snows were deeper and the summers sunnier when I was a youngster' or 'the seasons were more regular when I was growing up' are typical of elderly cohorts in nearly all societies when in conversation with younger generations. The ubiquity of such claims reveals something very important about the relationship between climate and the human mind. These memories of past weather shape how people expect the atmosphere to perform; the remembered past is a normative guide to the future. Winters *should* be snowier than they are today; summers *should* be sunnier; the rhythm of weather *should* follow a more discernible seasonal cycle. The recollection of past weather exerts powerful influences on the interpretation of present weather and on apprehensions about future changes in climate. Knowing climate through unstructured assemblages of remembered weather perhaps explains why many people are reluctant to embrace the prospect of novel climates, new assemblages of weather. People's idea of climate is predominantly shaped by the past, using it reassuringly to connect present and remembered experiences of weather.

(Sourced from: Hulme, 2016c)

Through empirical work conducted with rural citizens in Cumbria in England, Georgina Endfield and her colleague Alex Hall revealed the complex relationships that exist between local weather, community identity and cultural memory (Hall and Endfield, 2016). They studied people's memories of severe snowy winter weather and showed how personal knowledge of climate is moulded by local memories, stories and folklore (see **Figure 3.1**). These cultural resources are used to normalise severe weather and to sustain the community through future encounters with extreme climatic conditions. These attributes of human 'weathering' seem to transcend particular historical or cultural settings. For example, in a study of the Mande culture of Mali in Sahelian West Africa, anthropologist Rod McIntosh has shown how traces of historical climate-society interactions live on in the cultural memories of living descendants; what the Mande call *Kuma Koro*, 'the ancient speech of the ancestors motiving the living and their moral actions' (McIntosh, 2015: 275). For the Mande, long-past changes in climate are remembered through oral history – 'deep-time memory' – that reveal people's past emotional and imaginative interactions with climate. Not only are landscapes weathered by climates of the past, but so too are cultural memories and practices.

## Indigenous Climate Knowledge

As indicated in the previous section, personal knowledge of climate is always shaped and influenced by the wider culture in which a person lives or of which they have experience. This culture conditions how human memories of weather

**Figure 3.1**    (Left) The front and back of the original National Snow Survey postcards used in the late 1930s and (right) the front and back of the postcards distributed as part of the more recent 'Snow Scenes' project (Source: Hall and Endfield, 2016).

Reproduced by permission of Durham University Library

are shaped and recalled as well as the communal stories which normalise past experiences of extreme climate. It also conditions what evidence counts as warranting a specific knowledge claim: a standardised meteorological reading, the experience of a hurricane, a cultural narrative of fortitude or rescue in the face of climatic adversity. How such evidence is interpreted to construct authoritative accounts of local climate varies from culture to culture. Although there is a lively debate about whether 'indigenous' knowledge is in any meaningful way different from 'local', 'lay' or 'scientific' knowledges (Agarwal, 1995; Brace and Geoghegan, 2011), my point of using this category here is a simple one. I wish merely to emphasise that different sets of knowledge practices in different settings are used to construct authoritative knowledge about climate.

For many cultures today and in the past the relational worlds of meaning created through language and cultural practice are the starting point for how experiences of weather evolve into knowledge about climate. Huber and Pedersen (1997), for example, offer insight into such knowledge originating from Tibet. Here, in contrast to scientific practices of materialist reductionism, weather – and hence climate – is understood as a manifestation of the relationships between humans and spirit powers. In traditional Tibetan culture, physical encounters with the turbulent flows and conditions of the atmosphere are made sense of – are 'known' – through understanding the social relations between, and the respective moral duties of, humans and deities. This is similar to examples observed in various Pacific cultures in which knowledge of climatic fluctuations is deemed to reside in 'the domain of the gods' (Donner, 2007). Here, knowing climate and interpreting its behaviour is synonymous with knowing how one should act in the world of sentient human beings, animate spirits and divine agents.

A similar system for establishing reliable knowledge about climate has been described by Julie Cruikshank. Cruikshank (2001) explains how for the Tlingit of northwestern North America the behaviour of glaciers is determined not simply by physical climate, but also by human social relations. In the absence of written documents, oral traditions of glacier movements offer narratives in which the agency of humans and the agency of climate become entangled. Cruikshank also explores how such local climatic knowledge engages with knowledge emerging from the geophysical sciences. Although appearing incommensurable, she suggests that the cultural practices and metaphors of science may not be so different from those used and understood by the Tlingit. One such scientific practice is that of arduous and dedicated fieldwork – in which scientists (cf. Tlingit) take their eyes and their instruments out of the laboratory (cf. homes) into places where direct encounters with the physical world are possible (cf. glaciers). In this respect at least, one might argue that the Tlingit too are engaged in a process of scientific knowledge construction about their climate.

This blurring of distinct knowledge categories can also be seen in the example of indigenous climate knowledge-making in the Norwegian mountains. The Norwegian Public Road Administration often rely upon locally based snow clearers to issue warnings of impending avalanche risk for exposed highways during the winter season. These observers – often elderly and sometimes referred to as 'the snowmen' – have long experience of weather-watching and connecting their local climatic knowledge with avalanche risk assessment (Solli and Ryghaug, 2014). For the Norwegian snowmen, as with Cruikshank's Tlingit, careful observations are interpreted by local and historical experience in a process of 'assembling' climate knowledge. The authors of the study describe it thus:

> This knowledge practice involved a process of making sense of experience-based and often inherited knowledge in relation to interpretations of historic, present and predicted weather data and events. (Solli and Ryghaug, 2014: 21)

Nevertheless, questions remain about how different forms of climate knowledge engage and interact with each other in order to offer a basis for acting in the world. Tim Leduc (2007) offers a different view to Cruikshank about the commensurability of indigenous and scientific climate knowledge. He points out that while the Inuit term *Sila* is often interpreted as meaning 'weather' or 'climate', its local meanings are much broader and richer than understood by those English words. Leduc is therefore less optimistic than Cruikshank about finding 'cross-cultural research bridges' that would bring Inuit and western scientific climate knowledges into fruitful dialogue. Leduc's other important claim is that how these different knowledges of climate relate to each other cannot be separated from the legacy of colonial power relations which inflect Arctic cultures today. Indigenous knowledge of climate is not merely cultural, it also bears the mark of historical and political conflict and struggle.

## Scientific Climate Knowledge

I have already shown in **Chapter 2** how the emergence of scientific knowledge during the eighteenth century began to change the ways in which climate was understood in the western world. Standardised scientific practices of observation, measurement and tabular reduction allowed climates to be known through numbers for the first time. This enumeration of climate was to lead to the later establishment of statistical climatology in which climates across time and space could be precisely compared. Differences between climates in different locations were analysed numerically, while climate measured in one time interval could be contrasted formally with that in another. The possibility of climate *change* could now not just be imagined, it could also be quantified. One of the first people to do this systematically across large regions was the German geographer Eduard Brückner. His study at the end of the nineteenth century of fluctuations in climatic data from across Europe, Asia and America led him to propose, for the first time, the idea of the climatic normal – defining climate statistically on the basis of a fixed record length of standardised weather measurements, in his case 35 years (Brückner, 1890). This expanding enumeration of the weather in multiple locations allowed the idea of climate to be detached from the sensory human experience of weather and from the material and imaginative attachments of climate to specific places. Following Brückner, the operational scientific definition of climate became a statistical description in terms of the mean and variability of relevant meteorological quantities calculated over a period of time ranging from a few decades to (eventually) millennia.

Until the latter decades of the twentieth century, scientific knowledge of climate remained rooted in this statistical, largely descriptive, tradition. Other scientific field practices – for example, glaciology, dynamical meteorology, oceanography – each developed new methods of close observation and standardised instrumentation: ice coring, balloon ascents, weather buoys. Each discipline, in different ways, yielded new forms of climatic knowledge. One of these new field scientists interested in the scientific study of climate was the Swedish glaciologist Hans Ahlmann. Ahlmann used field science in the 1940s to establish knowledge of historical changes in climate in the high Arctic and constructed his theory of 'polar warming' using field observations of melting glaciers and through collaboration with local residents. Science historian Sverker Sörlin describes Ahlmann's observations as 'bodily knowledge' (Sörlin, 2011): knowledge made by the scientist immersing himself in a harsh material environment. It is similar, one might say, to the indigenous knowledge gained by the Tlingit. Sörlin's story explains how Ahlmann's knowledge was hard-earned through his close observation of challenging environments. But it also explains how this knowledge gave way during the 1960s to climatic knowledge emerging from other forms of scientific culture, namely satellite observation of the atmosphere and the development of climate simulation models.

These moves in the 1960s and 1970s towards an understanding of climate as a physical *global* system described using mathematical models demonstrates (again) how knowledge about climate is always culturally shaped. The new Earth observation satellites and computer models which inspired this globalisation of climatic knowledge were the outcome of post-war American technological and economic acceleration, fuelled in part by superpower rivalry with the Soviet Union. What then emerged in the 1980s were global coupled ocean-atmosphere models that became central for the emerging science of the Earth system (Edwards, 2010). For most scientists these models came to be seen as the dominant basis for climatic understanding and predictive knowledge (see **Chapter 9**), but their rapid evolution and rise to power can only be understood in the context of Cold War politics and the revolutionary technological cultures of the supercomputer.

By the latter decades of the twentieth century, for the scientific enterprise the idea of climate had become firmly attached to understanding the physical processes of an increasingly intricate and interconnected planetary system, as in the IPCC's formal definition cited earlier in **Chapter 1**. It is a concept of climate significantly different from the statistical concepts developed by nineteenth-century geographers such as Brückner. It is even further removed from place-based or community-shaped accounts of 'patterns of remembered weather in places', which I described earlier in this chapter. Rather than climate being understood by someone and from somewhere, this scientific conception of a systemic and mathematically simulated *global* climate offered a view of climate for everyone but from nowhere. Climatic knowledge became global knowledge. Global models, global temperature and global teleconnections became the vocabulary through which climate was to be understood scientifically. These 'global kinds of knowledge' (Hulme, 2010) erased geographical and cultural differences; the scale of meaningful climatic analysis was henceforth to be global.

Making such scientific knowledge about climate and its changes, as with other forms of climate knowledge, is hard-earned and culturally specific. It involves significant constellations of individuals, institutions and instruments sharing broadly similar cultural norms (see **Box 3.2**). The resulting epistemic culture of climate modelling contrasts markedly with Ahlmann's 'bodily knowledge' of polar warming, with the Mande's deep-time memory of local West African climates or with place-dwellers' personal knowledge of the climate of their farm or garden. Although the high social status of science in many countries and cultures begins to explain why the idea of *global* climate has become hegemonic, it does not lead to public knowledge of climate which is necessarily unitary and homogenous. The different knowledges of climate I have described above, and more varieties beside, circulate simultaneously in any given society, sometimes converging, sometimes not. But this observation leads to an obvious question: is there one climate, or are there many?

---

## Box 3.2: How Climate Models Gain Authority

The formalisation of knowledge required by climate models offers a powerful way – perhaps the single most powerful way – for scientists to organise their knowledge about the physical Earth system. Computer models help understand the material interconnections between the different physical components that comprise climate and help identify key sensitivities within the system. To construct, maintain and use a climate model implies at least a minimal level of understanding of physical causation in the complex Earth system and an ability to re-create features of that reality in a numerical simulation. It also requires huge financial and material resources and stable institutional and international networks.

Climate model simulations must correlate in some way with the observable physical world. If they do not, then as much effort must be invested in understanding the behaviour of the climate model as in understanding the

*(Continued)*

*(Continued)*

behaviour of the physical Earth system. After all, in such an eventuality it is the model that is deficient in some respect, not reality. Climate modelling can therefore in many ways be seen as a behavioural science, a science which studies the behaviour of climate models. But climate models also need to be studied as objects which exert powerful social influence. Whether or not the public, and whether the politicians they elect or appoint, *should* trust climate models when they are used to prognosticate the far future requires an additional set of questions to be answered.

For climate models to gain the status of trustworthy sources of climate knowledge it is necessary that they be evaluated against the criteria of reliability, validity and replicability. But it is not enough for climate models only to faithfully simulate an observed climate or for all models (broadly) to agree with each other. To gain authority within forms of democratic society it is important that the networks and practices that construct climate models are subject to critical scrutiny. Climate models and their social networks of design and execution must be held accountable to broader sets of public norms and standards. These norms are socially constructed and will vary between cultures and nations. It is insufficient to assert that climate models possess universal and uniform authority simply on the basis of their epistemic status. Ultimately it is the ways in which such claims of epistemic authority are socially validated that yield greatest insight into how and why knowledge based on climate models gains and exercises authority in a given society.

(Sourced from Hulme, 2013)

## Consensual Knowledge

When the United Nations voted in 1988 to establish the IPCC, it passed a mandate for this new institution to assess existing scientific knowledge with a view to reaching a consensus about what was known about climate and its changes. The IPCC has since gone through five assessment cycles and is now engaged in its sixth. But the IPCC has increasingly been confronted with the reality that important knowledge about climate is held not just in scientific archives, but in personal, social and local modes of knowing – as I have shown in this chapter. Important knowledge about climate is curated by indigenous communities, but never enters into the scientific literature. Personal and local knowledge of climate, beyond that captured in scientific studies, clearly is important for how people form their beliefs about climate and its changes.

The IPCC therefore faces a dilemma. It seeks to produce a knowledge consensus about climate and its changes, but its knowledge-sourcing is limited to science alone. By emphasising the primacy of scientific knowledge it constructs a universal and consensual view of climate from nowhere. But science emerges from within its own very special culture which inevitably marginalises other forms of climatic knowledge. It therefore ends up offering a view from somewhere, and that somewhere is from within the places, institutions and practices of science. The wider questions raised by this dilemma are challenging. What is gained and what is lost in producing such consensual knowledge about climate and its changes? Should different forms of climatic knowledge be integrated into a single whole? Are these different knowledge systems commensurable?

These questions can be addressed theoretically using the tools of epistemology and science studies, as many scholars have done. Such studies emphasise the different worldviews and power relations which are embedded in different knowledge systems. But answers can also be sought empirically, by engaging with actual communities in particular settings where climate knowledge is made and used. A good example of this latter approach is a study of the forms of climatic knowledge held by rain-fed farmers in southern Uganda (Orlove et al., 2010). The authors identified four components of placed-based climatic knowledge held by these farmers: familiarity with seasonal weather patterns, local climatic indicators, close observations of weather events and information about the seasonal progression of weather from elsewhere in the region. These elements of local knowledge were observed to interact with externally generated scientific weather forecasts and climatic indicators. Orlove and colleagues argued for the possibility of hybrid or emergent forms of climate knowledge to be constructed through this process. But while arguing for a symmetric appreciation of the value of different categories of knowledge, they fell short of suggesting a unitary ontology of climate.

## Chapter Summary

In this chapter I have described four different categories of climate knowledge: personal, indigenous, scientific and consensual. Although the distinctions between them have some value, certainly from a heuristic perspective, it is important to recognise that these categories are also somewhat arbitrary. Many people's knowledge of climate weaves together elements of all four. A more important point to take from this chapter is that what knowledge is held about climate and its changes is always a view from somewhere.

How much trust someone places in a knowledge claim varies from person to person and from culture to culture. There are many who claim to be the special holders or gatekeepers of climatic knowledge: scientists, other validated experts, community elders, personal witnesses, lay citizens. Different claims to knowledge about climate always emerge from particular sets of cultural beliefs, norms and practices. But which of these claims are deemed trustworthy and held to carry public authority is not just a cultural question, it is also a deeply political one. Placed-based perceptions, experiences and memories of climate may well carry greater authority and influence in some cultural settings than claims emanating from anonymous scientists or remote government institutions. Peter Rudiak-Gould captures this tension well in his investigation into the different ways in which climate change can be said to be 'seen':

> Ultimately, I think, the ... debate [between scientific and lay knowledge] boils down to an old tension in democratic theory and practice, the dilemma between two seemingly unacceptable positions: that scientists ought to have authority to speak over citizens, and that citizens can be trusted to make informed decisions about complex technical issues of which they have little understanding. (Rudiak-Gould, 2013: 129)

In the next chapter I extend my investigation of different climate knowledges to consider one particularly important feature of all such knowledge: their different accounts of why climates change over time and why climates become disordered.

## Further Reading

Callison, C. (2014) *How Climate Change Comes to Matter: The Communal Life of Facts.* Padstow: Duke University Press.
Edwards, P.N. (2010) *A Vast Machine: Computer Models, Climate Data and the Politics of Global Warming.* Cambridge, MA: MIT Press.
Leduc, T.B. (2016) *A Canadian Climate of Mind: Passages From Fur to Energy and Beyond.* Montreal: McGill-Queen's University Press.
Rudiak-Gould, P. (2013) *Climate Change and Tradition in a Small Island State: The Rising Tide.* Abingdon: Routledge.

# 4

# Changing Climates

## Introduction

Noah's Flood is a salient account of what might be called a global climatic catastrophe: the flooding of the world and the near extinction of humanity. The Flood is present in the sacred texts of Jews, Christians and Muslims, but echoes of similar mythological accounts of the survival challenges which confronted nascent human societies are also found in many other early civilisational histories. At the beginning of the eighteenth century in early Enlightenment Europe, a number of scholars expounded on the causes and consequences of the biblical Flood. One of these, an Italian philosopher by the name of Antonio Vallisneri (1661–1730), was particularly interested in the effects of the Flood upon global climate. In a treatise published in 1721, *Dei corpi marini che sui monti si trovano [Of Marine Bodies Found in the Mountains]*, he developed a new account of how the Flood caused a sustained deterioration in climatic conditions around the world and of the consequences of this change in climate for humanity.

There are two elements of Vallisneri's treatise that particularly interest me. For him, the mark of the deluge, and its subsequent reconfiguring of global climate, was left not just on the face of the land. Vallisneri also believed that it reconfigured the physiology of human bodies through degrading human reproductive capacity. A change in worldwide climate therefore had not just environmental consequences, but enduring bodily ones also. But second, and more relevant to this chapter, it was clear to Vallisneri that this change in climate should not simply be attributed to God's capricious or purposeless action in the world. As the biblical account of Noah's Flood makes clear, it was egregious human sin which provoked God into action[1]. Vallisneri belaboured the point: the root cause of this climate disruption was human behaviour and according to his reformist theology God was

---

[1] The account in Genesis reads, 'God said to Noah, "I am going to put an end to all people, for the earth is filled with violence because of them. I am surely going to destroy both them and the earth"' (Genesis 6: 13; New International Version).

entitled to pass judgement on humanity. Wilful human transgression against the divine order had consequences, in this case a disordering of the world's climate leading to the diminution of human health and fertility. Lydia Barnett's study of Vallisneri's work emphasises the point. In this early modern Protestant theology, she writes, 'the function of divine justice [was understood] as the material rendering of the spiritual state of individuals in and on their physical bodies' (Barnett, 2015: 228).

Of course, neither Vallisneri nor his contemporaries offered a scientific account of exactly *how* God's agency altered the flows of heat, air and moisture around the planet to bring about such a dramatic climatic deterioration. They were neither interested, nor able, to offer a persuasive naturalistic explanation for what had happened. Nevertheless, Vallisneri's treatise offers a coherent account of the causes and consequences of global climatic change which resonated with a particular prevailing worldview. What is also striking is how recognisable in today's dominant narrative of anthropogenic climate-change are some of the same key elements of Vallisneri's account. The contemporary narrative runs something like this: through willful and unconstrained consumptive behaviour, humans are responsible for changing present and future climates around the world, the consequences of which are, in a general and collective sense, proportionate to the degree of human culpability. God's agency in Vallisneri's account is replaced by nature's agency in the contemporary account, but the central causal role played by humans in bringing about changes to global climate remains the same.

\* \* \*

In this chapter I examine different accounts of causation which have been developed at different times and by different cultures to make human sense of changes in physical climates. It is not just religious thought and practice, least of all Protestant Christianity, which has provided the cultural contexts which foster the idea that humans and their climates are in complex, yet symbiotic, relationship. As I will show in the rest of this chapter, cosmologies, political ideologies, social practices and scientific paradigms of knowledge all contribute to the rich cultural matrix in which theories of climatic change and causation have emerged, flourished and declined. Which accounts of climatic causation you find persuasive today inevitably reveals something of your own cosmology, your own ideology and belief system.

The chapter is structured in three sections. In the first section I explore supernatural accounts of climatic change, in the second I emphasise naturalistic accounts of change and in the third I elaborate differing understandings of human agency and climatic change. But as should be clear from the above introduction, and will become clearer still, such explanations of climatic changes are neither discrete nor exclusionary. For many people today, as in the past, these different 'ways of thinking' interact with and shape each other. And in all cases they reflect wider cultural beliefs and practices.

## Supernatural Causation

One of the central arguments of this book is that the prevalence and power of the idea of climate in human cultures is due to its role in stabilising relationships between changeable weather and cultural life. The idea of a stable climate is therefore readily associated with the idea of a stable cosmic order in which relationships between humans, non-humans and the spirits or gods are as they should be. For many cosmologies this suggests that disruption to any part of this triadic relationship may yield adverse consequences for the behaviour of the weather and challenge the human experience of a stable climate. For example, the Abrahamic faiths – namely Judaism, Christianity and Islam – understand a transcendent and omnipotent God to be the provider of all good things, including orderly and faithful weather as I showed briefly in **Chapter 2** in the case of Jehovah. More generally, if deities or spirits are powerful, awesome and just, then a prerequisite for retaining a beneficent climate is for humans to maintain good and appropriate relations with them. And if the deities are merely capricious, then various petitions, offerings and sacrifices are needed to appease them, thereby maintaining orderly weather around which human life can at least survive, if not flourish.

Within such worldviews it is both normal and sensible to search for the hand and motives of a good and just God when climates start to change or to acknowledge the anger of spirits when weather appears to become abnormally extreme or destructive. Fears and anxieties about extreme weather and dangerous climatic shifts can be made sense of, if not fully diffused, by placing responsibility for the performance of climate in the will of the divine (this may be true whether or not accompanying naturalistic explanations for changing physical climates can be offered; see below). Such an ideology creates powerful and binding narratives about the performance of wayward climates. These narratives contribute to psychological and spiritual survival, even as all around might be physically threatened.

Explanations for changes in climate such as these are frequently found in historical cultures and remain prevalent in the world today. European societies in the sixteenth and seventeenth centuries, for example, struggled to comprehend a climatic period which encompassed some very severe winters and cold, damp summers. Religious songs and prayers written at the time reveal the value of a theistic explanatory framework. The German hymn writer and poet Paul Gerhardt (1607–1676), for example, wrote 'Song of confession and prayer occasioned by great and unseasonable rain[2]' in the 1640s. In this extended poem Gerhardt looks to God for an understanding of the climatic perturbations he and his contemporaries in central Europe were then experiencing, and also for a resolution to their experience:

---

[2] [*Buß- und Betgesang bei unzeitiger Nässe und betrübtem Gewitter*]; trans. John Kelly 1867.

O God! Who dost Heav'ns' sceptre wield
What is it that now makes our field,
And everything that it doth bear,
Such sad and ruin'd aspect wear?

Ah! Father, Father, hear our cry,
Redeem us, 'neath sin's yoke we lie,
From out the world drawn may we be,
And Thou Thyself turn us to Thee.

Wrath's black robe tear off with Thy hand
And comfort Thou us and our land,
And may the genial sun shine forth
And ripen the fair fruits of Earth.

(Extract, taken from Rockoff and Meisch, 2015)

Other spiritual explanations of climatic misdemeanours may be more sinister, drawing attention to a Manichaean struggle between good and evil. On occasions and in some cultures the devil's work in disrupting the orderly state of climate has been accomplished through the weather-making witch. In sixteenth-century continental Europe, climatic deterioration was not uncommonly attributed to the unleashing of demonic forces through witchcraft and weather magic (see **Figure 4.1**). Misogynist elements of European society held witches directly responsible for a variety of social ills and 'unnatural' phenomena, not least the high frequency of damaging weather episodes, especially in winter. Although such claims of climatic causation were fiercely debated by theologians at the time, at its most extreme such misogyny could extend to witches being burned as scapegoats for the damage caused by a volatile climate (Behringer, 1999). This most perverse form of European weather regulation, through witch-hunts and burnings, occurred most commonly in the century from 1550 to 1650 CE. The practice was successfully challenged on occasions by the reformist Lutheran Church, insisting that it was neither the devil nor his nefarious agents who was responsible for the weather, but God alone. Naturalistic explanations for such climatic deviance also struggled for ascendancy during this period (see below).

Other cosmological systems similarly display beliefs which attribute climatic agency to supernatural beings. For example, the Tapeños – an indigenous people group in the Andean highland region of Peru – invoke mountain deities and ancestors to explain changes in their climate. Rituals and supplications to these spiritual agents are to secure timely and sufficient rains, or at least are offered to assuage these deities' anger at communal cultural decay which would otherwise manifest through a withholding of rain. Yet studies of this Andean culture also show how such indigenous beliefs and practices do not remain static (Paerregaard, 2013). As with the weakening of demonic explanations for

**Figure 4.1** Witches 'cause' a hailstorm. Illustration from the title page of Ulrich Molitor's *Of Witches and Diviner Women* (1489) (Source: Behringer, 1999).

climatic change in seventeenth-century Europe, the Tapeños have recently begun to assimilate naturalistic explanations for their climatic fortunes alongside their supernatural accounts of agency.

The more general point here is important. Physical climates change through time; but so too do theories of climatic causation. Explanatory accounts of why climates change do not remain static. As cultures evolve, often in response to experiences of environmental change, cross-cultural encounters, new scientific knowledge and technological innovation, so too do explanations of climatic change and variability.

## Natural Causation

Supernatural and natural theories of climatic change are not simply alternative discrete accounts which humans have found useful to explain the (usually) unwelcome inconstancy of climate. They rarely present as two mutually exclusive options from which people select. Neither is it possible to develop a chronological account of changing theories of agency, in which naturalistic explanations of climatic change have gradually squeezed and then finally supplanted supernatural ones. Both categories of explanation have co-existed for most of recorded history, even if their respective salience and cultural authority has varied over time. Thus in Classical Greece, Aristotle was an early champion of materialist explanations of physical phenomena, including the weather. And it was one of Aristotle's students, Theophrastus in the third century BCE, who left an early naturalistic account of climatic change: local warming of the climate around Philippi being attributed to the physical effects on the atmosphere of forest clearances.

Yet not only are there eras in Earth history when physical climates change rapidly. There are moments in human history when *ideas* of climate and its causes also change rapidly. In other words, there are particular times and places when new ideas about the world emerge and become creative, pervasive and culturally authoritative. With respect to theories of climatic change one such place and moment was western Europe in the nineteenth century. It was here, and it was then, that the novel idea that climates could change over vast epochs without God's direct agency became rapidly accepted across the western world (see **Box 4.1**). The conditions for such intellectual novelty were lain down in earlier centuries of European inquiry through the rise of empiricism, scientific instrumentation and global exploration. Also necessary for the emergence of such new theories of climatic change was a re-imagination of the European idea of time, aided by the work of eighteenth-century geologists such as James Hutton. Rather than 6,000 years of a divinely maintained climate, it became possible to imagine that the Earth may have manifest many large fluctuations in climate during a 'deep past' consisting of millions of years.

---

### Box 4.1: Climate and *Die Eiszeit* ('the Ice Age')

The conventional view of climate prevailing in Europe at the end of the eighteenth century was that of a young Earth and a stable climate. This divinely 'designed' physical state was believed to be punctuated by occasional natural catastrophes – earthquakes, floods, volcanic eruptions – events which could seemingly account for all of the emerging geological and fossil evidence of a dynamic Earth. The only agent capable of radically de-stabilising the world's climate remained God, for example, in the idea of the biblical Flood (see the opening of the chapter).

One of the first people to contemplate that powerful *natural* forces could compete with God in achieving similar scales of climatic change was the German philosopher-poet Johan Wolfgang von Goethe (1749–1832). By studying erratic boulders in the North German Plain, Goethe concluded that there must have been an 'epoch of great cold' when glaciers, the conveyors of these erratics, extended well to the north of the Alps. Goethe's evidence was later affirmed in the 1820s and 1830s through the detailed studies of a small group of Swiss scientists and engineers, motivated in their searches to confront the hazards of bursting ice-dammed Alpine lakes. They too became convinced that glaciers had once extended well beyond their then known limits, prompting the German geologist and poet Karl Schimper to coin the neologism Die Eiszeit, 'the Ice Age' in 1837 (Cameron, 1964).

But the great populariser of the Ice Age was the Swiss naturalist Jean Louis Rodolphe Agassiz. Taking inspiration from Goethe's imaginative leap, and the evidence gathered by his older Swiss colleagues, in communication with Schimper the ambitious Agassiz became convinced of the plausibility of such a hypothesis. In front of the Swiss Society of Natural Sciences in July 1837, he presented the first coherent glacial theory of climatic change and three years later published his widely circulated book *Études sur les Glaciers*. The idea that the Earth's climate, unaided by God, was susceptible to such large changes in climate, and over such long time-scales, was not an easy one for mid-nineteenth-century Europeans to accept. It sat uneasily alongside traditional biblical interpretations of a (fixed) created world and of the supposed beneficent hand of God preserving climatic conditions broadly suitable for human habitation. Yet by the 1870s, Agassiz's glacial theory of climatic change was widely established in scientific circles and in popular culture on both sides of the Atlantic. The world's physical climate had not changed significantly between 1780 and 1880, but the way people thought about climate and its changes had altered radically.

With the realisation of the vast natural forces required for global climate to vacillate through repeated glacial cycles, the search was on to identify the precise causal agents, beyond God, of such disruptions. The later nineteenth and early twentieth centuries witnessed competing ideas vying for ascendancy: irregularities in the Earth's orbit (James Croll, 1870s), changes in the atmosphere's gaseous composition (Svante Arrhenius, 1890s), solar cycles (Ellsworth Huntington, 1920s) and, later in the twentieth century, volcanic eruptions (Hubert Lamb, 1960s). But it was not just changes in millennial- and global-scale climates that required naturalistic theories of change. During the latter half of the twentieth century increasing attention was paid to understanding causes of climatic change on much shorter time-scales. Although earlier geographers such as Eduard Brückner and Ellsworth Huntington had been interested in such questions, physical understanding of inter-decadal changes in regional climates remained poorly developed. The British climatologist Hubert Lamb could therefore write in 1959:

> Not so very long ago – between the wars in fact [1920s–1930s] – climate was widely considered as something static, except on the geological scale, and authoritative works on the climates of various regions were written without the allusion to the possibility of change...
> (Lamb, 1959: 299)

After the Second World War, and especially after the International Geophysical Year of 1957/58, new global observing systems and computational machines contributed to refining naturalistic explanations for changes in climate.

It became possible to test scientific understanding of how natural interactions between different physical elements – oceans, ice-sheets, clouds and forests – generated climatic instabilities on multiple scales. Not least of these was the idea of El Niño, the vast coupling of the Pacific Ocean and atmosphere with tropical and extra-topical weather systems, which led to regional perturbations in climate every three to seven years. El Niño has since become emblematic of the scale and intensity of natural variability in regional climates, a phenomenon amenable to scientific investigation and prediction. But it has also increasingly taken on a role of offering an independent, almost quasi-cultural, explanation for all sorts of climatic ills and dangers. Yes, El Niño can be 'explained' through purely physical processes without invoking the supernatural or the human. And yet this peculiar personification of climate[3] offers a scapegoat for climatic misdemeanours around the world, in which agency can helpfully be attributed to an apparently comprehensible 'thing'. As one American media commentator noted in October 1997, following a particular powerful El Niño, 'We used to blame the Soviets [for extreme weather], now we can blame El Niño' (*Chicago Tribune*, quoted in Grove and Adamson, 2017: 213). I explore this question of climate and blame further in **Chapter 6**.

## Human Causation

The previous sections offer some insights into different cultural understandings of what causes physical climates to change. Supernatural entities may be believed to cause climates to change, and they may do so through either natural or supernatural means; i.e. with or, as in the biblical Flood, without accompanying naturalistic explanations. Climates may also change for entirely natural reasons, as in glacial cycles. Thus the three possible causal chains shown in **Figure 4.2** are accounted for: routes 1, 2 and 3. At best, however, this is only a partial description of how people think about the causes of climatic change. I have made it clear that human agency is implicated in diverse and complex ways in most supernatural accounts of climatic causation, and I will now show that similar complexities abound when thinking about human agency and natural causation.

Of far greater interest and concern for most cultures has been to understand the place of human agency in these supernatural and natural chains of explanation. Does God act independently of human behaviour? Is nature unresponsive to human actions? The answer to both these questions, at least for most people in most cultures and in most eras, has been 'no'. A more

---

[3] El Niño, or 'the little boy', was originally named by Peruvian fishers to identify the annual appearance of warm water offshore which occurred in the run-up to Christmas. Grove and Adamson (2017) explain how this nomenclature evolved during the twentieth century to become this quasi-human personification of a climatic phenomenon.

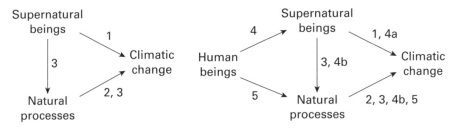

**Figure 4.2** Schematic representation of causal reasoning accounting for changes in climate: (left) without human agency; (right) with human agency. See text for explanation of the five causal 'routes'.

complete schema to capture the range of possibilities therefore is that shown on the right side of **Figure 4.2**. The question then becomes, 'How do different cultures, and different people within particular cultures, apportion responsibility for climatic change between nature, their gods and themselves?' The boundaries between these different modes of explanation are far from clear, are never static and are frequently contested.

It seems likely that the idea that humans can and do cause climates to change is as old as the idea of climate itself (von Storch and Stehr, 2006). Aristotle and his disciples believed that human-cleared forests caused the climate of Philippi to warm. Vallisneri and his Protestant colleagues believed that it was human wickedness which provoked God to intervene to cause the Flood. And in post-revolutionary France in the early nineteenth century, the socialist Charles Fourier was convinced there was a decline in the health of planetary climate caused by human greed. For Fourier, economic motives and rampant individualism had led to widespread land clearances, hence his claim that 'Climatic disorder is a vice inherent to civilised cultures that disrupts everything due to the battle between individual and the collective interest' (Fourier, 1822; cited in Locher and Fressoz, 2012: 587).

One should be wary of a presumptive exceptionalism which thinks that it is only late-modern westernised cultures which have identified a role for human agency in causing climates to change. Of course, through the enterprise of science, they have indeed done just that, hence the consensus claim of the United Nations' IPCC: 'Warming of the climate system is unequivocal [and] human influence on the climate system is clear' (IPCC, 2013). Today, a majority belief is that human activities alter the physical composition of the atmosphere which, through natural processes, causes world climates to change (i.e., route 5 on the right side of **Figure 4.2**). And some would go further. Not only are human activities causing climates to change, but collectively these activities have sufficient reach to change all aspects of the planet's functioning, through notions such as 'tipping points' (see **Box 4.2**). This bigger claim is often linked to the idea of a new geological epoch called 'the Anthropocene', the age of humans (see **Box 12.1**).

# Box 4.2: Abrupt Climatic Change and 'Tipping Points'

Naturalistic understandings of climate continue to evolve. As they do, they offer different cultural resources for thinking about the relationship between climatic change and human agency. One example of this creative role of the scientific imagination in cultural life is the idea of tipping points. In scientific terms a climatic tipping point might be thought of as occurring when a climate system crosses some physical threshold after which some change in that system becomes abrupt and irreversible. But a tipping point is also a metaphor, a figure of speech which has communicative value. It helps audiences grasp an unfamiliar idea, namely abrupt climatic change, through deploying a more familiar idea, for example, that of an object (like a balance) suddenly moving from one state to another.

The idea of rapid changes in climate was originally promoted by the American oceanographer Wallace ('Wally') Broecker in the 1980s, building upon new evidence extracted from cores drilled into the Greenland ice-sheet. Broecker connected the evidence for relatively rapid changes in climate in the past with the possibility that human perturbations to climate in the future might also trigger abrupt changes in aspects of the Earth's climate. His 1987 article in the journal Nature was titled 'Unpleasant surprises in the greenhouse?' He suggested that rather than thinking of human-caused climate-change as being gradual, it should be recognised that the Earth's climate might respond to human forcing in sharp jumps: 'I fear that the effects of [global warming] will come largely as surprises' (Broecker, 1987: 126). This new thinking about climatic change was to lead scientists to find new ways of conceiving, representing and modelling climatic change. Ideas such as thresholds, abrupt and non-linear changes and, from 2005 onwards, tipping points became part of the new paradigm of Earth system science.

As with all metaphors, climate tipping points open up new possibilities for the human imagination to work with. Broecker himself was able to harness the creative power of this new line of thought when he described human agency thus, 'The climate system is an angry beast and we are poking it with sticks' (quoted in Stevens, 1998). In recent years tipping points have been given expression in diverse cultural and political forms. Thus new art work and advocacy movements have appropriated the tipping point metaphor in creative ways, as in the arts organisation 'tippingpoint' with its strapline, 'energising the creative response to climate-change'[4]. And in public discourse evoking climate tipping points carries great rhetorical power, as in UK Prime Minister Tony Blair's claim from 2006 that 'we have a window of only 10–15 years to take the steps we need to avoid crossing catastrophic [climate] tipping points' (quoted in Watt, 2006). How climate is conceived and communicated scientifically changes the imaginative scope for cultural and political action in the world.

---

[4] www.tippingpoint.org.uk (accessed 11 May 2016).

The elevation of human agency to the status of a geological force is no uniform or universal belief. It interacts in multiple ways with other spiritual and supernatural beliefs to form more complex understandings of human agency with respect to climate. Inhabitants of the Marshall Islands in the Pacific believe that global warming is caused not just by industrial emissions from the Global North, but also by a local decline in cultural integrity and social order, a syncretist explanation for climatic change that Peter Rudiak-Gould (2012) has called 'promiscuous corroboration'. For the Porgeran people of Papua New Guinea climate change is due to the ineffectiveness of rituals oriented towards powerful spirits that control the cosmos, while for the Anglican Bishop of Carlisle flooding episodes in England in 2007 were a judgement on the greed and immorality of modern society. In similar vein, theologian Michael Northcott calls global warming 'the Earth's judgement on the global market Empire and the heedless consumption it fosters' (Northcott, 2007: 7). These voices offer accounts of climatic causation which follow routes 4 and 5 on the right side of **Figure 4.2**.

It seems then that human fingerprints on the world's climate are understood to be found in many different ways and places. Human agency is everywhere even if it is unevenly distributed; not everyone is equally to be blamed for a change in climate. Culturally diverse as these accounts of agency are, the one idea which appears untenable in the contemporary world – indeed, if ever it was – is that changes in climate exist independently of human thought and action. However, a new cohort of climate scientists is embarking on a quest to differentiate forensically between extreme weather which has been made by humans and extreme weather which occurs free of human influence (Hulme, 2014b). This new science, called 'probabilistic event attribution' (PEA), seeks to isolate what (and by implication who) lies behind each climatic event in the world, or at least the most extreme ones. The quest is charged with political, legal and ethical significance.

PEA aims to discriminate, scientifically, between two categories of extreme weather: 'tough luck' or 'human-caused'. People subject to 'tough luck' weather are victims (or occasionally beneficiaries) who cannot hold other humans to account for their meteorological fate. They must blame their weather either on 'Acts of God' – the hollowed-out theological explanation of extreme weather still used at times by the insurance industry – or else on the random workings of a morally blind nature. On the other hand, people subject to 'human-caused' weather are victims (or occasionally beneficiaries) who now have been offered by PEA scientists the possibility of holding a morally responsible human agent to account. *Their* weather is attributed to the accumulation of greenhouse gases in the global atmosphere. These emissions result from wilful human actions, whether those of individual high-emitters ('carbon sinners') or corporate actors responsible for supplying fossil-carbon energy sources in the first place ('carbon dealers'). A litigious pathway for receiving compensation for the experience of 'bad weather' is thus opened. Under this scenario it is not God, witches or the devil who are excoriated for deleterious changes in climate, it is human carbon sinners and nefarious fossil fuel corporations who are placed in the dock.

## Chapter Summary

Human anxieties about a disorderly climate are long-standing. They manifest today, for example, in the popular descriptions of climatic change as 'weather weirding' or 'climate chaos'. If, as I am claiming, climate is an idea which performs important functions in stabilising relationships between the experience of weather and cultural life, when physical climates appear to change the search for explanation becomes pressing. Have the deities been provoked to punish a wayward humanity? Is an offended God calling for repentance? Has the atmosphere been 'prodded and poked' too far by excessive human behaviour for it to be able to provide the climate people desire? In this chapter I have examined three modes of explanation for changes in climate: the supernatural, the natural and the human. I have argued that these explanations co-exist in complex ways within and across different cultures and that there is an ebb and flow to their respective cultural authorities. What is also clear is that it is exceptional for humans to think that climates, at least the climates of the human past and the human future, change for either natural or supernatural reasons *alone*. It is far more common, and indeed perhaps more necessary, to believe that the performance of climate is tied to the behaviours of morally accountable human actors.

In his 2013 Holberg Prize Lecture 'Which language shall we speak with Gaia?', the anthropologist Bruno Latour asked where agency for climate resides in the world. Who or what is to take responsibility for its changing behaviour? Latour claimed the political task ahead was to 'distribute agency as far and in as differentiated a way as possible' (Latour, 2014: 18). It is to recognise that human beings are neither pure subjects, mastering climate for their pleasure, one might say, nor pure objects, being mastered by climate to their detriment. Or, put differently, the political task is to recognise that climate and human life have historically been understood not as two separate domains with one causing or shaping the other. Rather, for much of the past and in most places, climate and humans have been understood to move together, their agency and fate conjoined through the mediating roles of natural processes and supernatural beings (**Figure 4.2**). Although the underlying idea of human responsibility for climate is not new, today's hegemonic discourse of anthropogenic climate-change perhaps makes it possible – at least for those living with the legacy of the western enlightenment – to see this unavoidable intimacy more clearly. I explore some of the implications of this co-dependency for the future of climate in **Chapter 12**.

\*\*\*

The following four chapters constitute Part 2 of the book, 'The Powers of Climate': what range of outcomes does the idea of climate accomplish in cultural life? I explore how climates and cultures interact in specific places to shape modes and patterns of everyday behaviour (**Chapter 5**) and how the idea of climate animates different cultural accounts of blame (**Chapter 6**) and fear (**Chapter 7**). I also consider the power of climate to provoke the human imagination and the many ways in which climate and its changes are represented in cultural form (**Chapter 8**).

## Further Reading

Grove, R.H. and Adamson, G.C.D. (2017, in press) *El Niño in World History*. London: Palgrave MacMillan.

Fleming, J.R. (2005) *Historical Perspectives on Climate Change* (2nd edition). Oxford: Oxford University Press.

Parker, G. (2013) *Global Crisis: War, Climate Change and Catastrophe in the Seventeenth Century*. New Haven, CT: Yale University Press.

Woodward, J. (2014) *The Ice Age: A Very Short Introduction*. Oxford: Oxford University Press.

# Part 2
## The Powers of Climate

# 5

# Living with Climate

## Introduction

It is easy to demonstrate that there is no universal 'ideal' climate. Ask any random group of people in your neighbourhood and you will find diverse and often contradictory descriptions of their favoured climate. This diversity will express itself in different preferences for year-to-year reliability, seasonal contrasts, levels of sun and cloud, gentle rain versus dramatic storms, and so on. But it will perhaps be most noticeably expressed in terms of thermal ideals: at what temperature does someone feel most comfortable? Research from around the world has shown that reported outdoor thermal ideals range widely, from 6 to 30°C (Shove, 2003). Differences in preferred thermal climates arise for many reasons of course: human physiology and age, clothing conventions, amount of time spent outdoors, gender, wealth, different activity patterns and lifestyles. My point nevertheless is simple: how people live with their climate – the adjustments made to make that experience more tolerable or productive – is diverse.

These differences in the lived experience of climate are brought into acute focus when one contrasts the ways in which people acclimatise to indoor and outdoor climates. Again, just ask any random group of people – even one drawn from within an extended family unit – and you will find different preferences expressed about the temperature at which a room or car thermostat should be set. This diversity will be for many of the same reasons as mentioned above. Yet as the world's population becomes increasingly urbanised, more people are spending greater amounts of time in climatically controlled indoor environments: in homes, offices, vehicles, cinemas and restaurants. Western people now spend around 90 per cent of their time in buildings or covered spaces whose micro-climates can be controlled through smart architectural designs and/or through mechanical devices. People's experiences of outdoor and indoor climates are increasingly diverging. Outdoor seasonal rhythms, and the daily flows of unregulated weather that accompany them, become detached from the (often) heavily regulated indoor climates. In **Chapter 10** I examine the technologies and discourses of *global* climate control, aka climate engineering, but it is important to recognise that most people's experience of climate control lies much closer to home, literally.

One way of understanding the decoupling of outdoor climates and thermally regulated indoor spaces is to follow the daily practices and habits of people in different settings – i.e. their acclimatisation routines. The geographer Russell Hitchings did this for a group of professional London office workers. He found that they were largely insulated from the passing seasons because of a near-constant office thermal climate and because of dress codes and social rituals which no longer responded to the changing weather outside. Describing this retreat indoors as like being '… cocooned in mechanically conditioned air', Hitchings captured the sense of detachment through the remarks of one of his respondents. Reflecting on her inability to notice the passing of London's summer climate, this respondent commented:

> I don't think I would have noticed the lack of summer otherwise. It's disappointing when you actually think about it, but most of the time I was only thinking about it because someone on the news was talking about there not being any summer and telling me what that means. (Hitchings, 2010: 289)

Social practices, cultural norms and material technologies condition the human experience of climate, whether for office professionals in London's financial district or rice farmers in the northern hills of Thailand. These practices, norms and technologies are many and varied, for example, showering and bathing regimes, public and private dress codes, single- or triple-glazed windows. They are also in constant flux. *Climates* change, whether indoor or outdoor, because cultures change. Just think of the rise of indoor central heating in nineteenth-century America (Meyer, 2000). But the inverse is also true: *cultures* change because climates, whether indoor or outdoor, change. Just think of the growth of artificial snow resorts in recent years. 'Human beings live culturally', as anthropologist Mary Douglas observed, but they also live climatically, amidst the patterns and fluxes of weather that they encounter in different places. The human experience of climate, both material and imagined, conditions cultural ways of life. Climates and cultures exist in dyadic relationship.

\* \* \*

In this chapter I illustrate how weather and culture interact in everyday life in specific places. This lived experience gives form to the idea of climate and imbues it with personal and collective meaning. Climates are therefore 'lived' through, *inter alia*, the atmosphere, landscapes, social imaginaries, clothing, built environments, emotions, ritual and weather-talk of specific places and cultures. Collectively, these ways of living climatically may be thought of as 'weather-ways' (de Vet, 2014), the patterned expectations, experiences and responses to daily weather among ordinary people. Bodies, cultural pursuits, material artefacts and memories (see **Box 3.1**) respond to the weather to which they are exposed. The idea of climate becomes inseparable from imaginative, social and material practices. People and their cultures become 'weathered'.

The chapter proceeds in four sections. First, I examine the ways in which ideas of climate, place and identity are tightly bound together through the experience of weather and how this manifests in different social practices and cultural norms. I then reflect on how these ways of living with climate are sensitive to change: first, to changes in physical climate, then to changes in culture and then, finally, to simultaneous changes in climate *and* culture. How humans live with their climates is in continual flux, driven by human mobility, by cultural innovation and by changes in the weather itself.

## Climates in Place

A central argument of this book is that the idea of climate emerges at the interaction between the human experience of weather and cultural practice. It is therefore necessary to ground any investigation into how people 'live with' climate in specific places. Making sense of climate and its changes cannot be separated from how weather enwraps itself with landscapes, memory, the body, the imagination and routine practices in particular places. Approaching climate this way demands an explicitly geographical and cultural interrogation of how people live climatically, how they become weathered. Climates are made personal and meaningful to people inescapably in and through places. This is one of the reasons why many regard the idea of 'global climate' as imaginatively impoverished (Ursula Heise's *Sense of Place and Sense of Planet* (2008) is a thorough exploration of the difficulties in establishing imaginative attachments to a global climate). Abstract and disembodied scientific claims about a changing global climate have no correlates in lived human experience. In the words of historian Vladimir Janković, '... [humans] simply do not *meaningfully* reside in such an entity [as global climate]' (2009: 178; emphasis in the original). On the contrary, humans *do* reside meaningfully in places.

In contrast to climate scientists, anthropologists are interested in what the idea of climate means to people *in places*. Fruitful lines of approach for such study are the ideas of 'weather-worlds' (Ingold, 2007) and the processes of 'weathering' (Vannini et al., 2012). Ingold's weather-worlds are imaginative constructs which tie together the felt experience of humans living 'in the open' – in the atmosphere with its incessant stream of weather – with the textures of the land they inhabit. Just as landscape theorists recognise that the surface of the land carries layers of cultural meaning, so too the atmosphere – and the weather it yields – is rich in meaning. Many of these cultural meanings are bound up with memories of past weather events. These memories may often be inscribed, physically or imaginatively, on local landscapes as a result of past floods, droughts or rare snowfalls. They may also be reified in visual culture through paintings or photographs of physical markers of past weather, or memorialised through monuments. These material and cultural encounters with past weather shape the linguistic repertoires through which the climate of a place is remembered and recounted by a community. The canonical stories which arise describe the character of local climates far more powerfully than do abstract meteorological statistics. For example, in my own past I remember the hot summer of 1976 in England (see **Box 3.1**), not because of the record thermometer readings of that season, but because of the remembered cricket matches, barbecues, camping and sea-bathing which the sun and heat enabled.

Weather-worlds are therefore always place- and person-centred. They emphasise how the idea of climate becomes constructed imaginatively, how it becomes inseparable from my sense of place and of self and from my lived daily routines. It is not just inanimate entities such as trees and buildings that become weathered through physical and chemical processes of change and decay. People and places also 'weather' with time. Their personal identities, social practices, material technologies and cultural memories become shaped by the atmosphere until, gradually and imperceptibly, these people and places embody the weather to which they are exposed – they become weathered. Yet this is not a passive surrender to the physical forces of the atmosphere; I am more than a direct function of the weather I experience. I may not be able to change the physical weather in my outdoor atmosphere, but I *can* change how I imagine the weather and, within limits, how I choose to live with it. This agency manifests in the everyday decisions of my personal and social life (see **Box 5.1**). It also manifests in the inner life of my emotions. In her study *Weather Reports You*, the artist/anthropologist Roni Horn captured the richness of this active psychological engagement with climate in her informal survey of hundreds of ordinary Icelanders living in the vicinity of the town of Stykkishólmur in western Iceland (Horn, 2007). Although exposed to a near-identical physical climate in this rather small region, participants displayed a wide range of emotional engagements with their weather: fear, hope, beauty, sadness, nostalgia, anxiety, joy, freedom. Social practices and emotional responses together construct the weather-worlds which give shape to the human idea of climate.

---

### Box 5.1: Weather in Everyday Life

For many people the idea of climate is bound up with the everyday things that they can or cannot do because of the weather: when is it warm enough to swim outside; when can rice-seeds be planted; when are winter tyres needed for the car? Different societies, and different people within a society, will therefore apprehend their climate very differently according to their livelihood, their social status, their access to specific technologies and so on. Not least in this regard is the increasing urbanisation of the human population and how 'the city' influences everyday weather practices. With well over 50 per cent of people now living in cities, how climate is experienced is changing for many people.

In a study of the daily lives of Australian citizens in the cities of Darwin and Melbourne, the geographer Eliza de Vet observed carefully how people experienced their weather and how they responded to it by adjusting daily routines and practices. These 'weather-ways' were the patterned expectations, experiences and responses to daily weather among ordinary people which for them constituted an embodied reality of the abstract idea of climate. Although people's physical experience of weather in a place may be similar, their understanding of a place's 'climate' will be conditioned, at least in part, through such material and cultural practices. What does this climate in this place allow me to do or prevent me from doing? What risks does it present me with? How are my moods affected?

In these two contrasting cities – Darwin in the tropics and Melbourne in the mid-latitudes – de Vet was particularly interested in how people experienced thermal comfort. She found that perceived climatic comfort was not just a function of degrees of temperature, but also of the duration of heat or cold and of the interaction between temperature and other weather elements such as humidity, wind, cloud and rain. But further than this, personal circumstances (e.g. office worker vs. home worker) and active pursuits (e.g. outdoor recreation vs. indoor leisure) greatly affected people's material and imaginative engagement with their climate. Whether people coped with thermal discomfort by changing clothing, switching on mechanical devices, moving to a different indoor (or outdoor) space, altering drinking and eating habits, or using alternate modes of transport and so on, could be understood only with respect to these individual contexts.

Wider cultural norms and access to technological devices were also important in the everyday experience of climate. Thus dress codes shaped and constrained the ways in which Australians coped with different thermal extremes and the nature and availability of air-conditioning (in the home, car or office) interacted with personal levels of thermal tolerance. Indeed, de Vet found that different expressions of physical and mental tolerance were an important aspect of acclimatisation in both cities. Tolerance was an accepted strategy because participants conceded that occasional discomfort was part of what it was to live climatically. Understanding the lived experience of climate in this way, through everyday social practices, yields rich insight into the range of adaptive practices that people learn, exercise and evolve.

(Sourced from de Vet, 2014)

Human weathering therefore has both material and immaterial dimensions, something explored by Phillip Vannini and his colleagues in the rainy climates of British Columbia's Pacific coastal regions (Vannini et al., 2012). They showed how shared weather experiences, and the embodied practices that flow from them, form bonds between people and the landscapes they inhabit. The climate of British Columbia's coastal regions is not just some numerical correlate of the patterned weather of the sky. The climate of this place as understood by its inhabitants emerges through an ensemble of memories, social practices, shared identities, material artefacts and emotional states. One cannot understand the idea of climate without understanding the material and imaginative attachments to place brought about by the experience of weather. This sense of climate is nicely captured in the remark attributed to the comedian Karl Valentin in respect of the climate of Munich, in southern Germany: 'I always feel the föhn [a dry, warm, down-slope wind off the Alps], even if it's not there' (cited in Lüdecke, 2009: 216). This holistic apprehension of climate is more intuitive than the scientific understanding of climate; it certainly remains widespread in today's world.

This lived experience of climate also yields insight into how the moral valence of local weather is constructed. Expectations of climatic normality and abnormality influence what types of weather are deemed appropriate for the time of year. What constitutes 'good' or 'bad' weather is conditioned by cultural identity, social practices, economic structures and social status. This can be seen in the case of Bergen, a city in western Norway. Elisabeth Meze-Hausken (2007) studied the construction of good and bad weather by following the climate reporting in the region's daily newspaper, *Bergens Tidende*. Bergen is the wettest city in Norway and many of these media stories revolved around rain and, its opposite, sunshine (see **Figure 5.1**). Yet it was not the amount of rainfall *per se* that created the headlines. Rather it was the context of the surrounding event that mattered, whether it was damage or accidents caused by weather calamities, the weather occurring on culturally significant days and festivals, stories which conveyed a poetic or romantic view of the weather or reflections about whether Bergen's climate was living up to the city's own, culturally conditioned, expectations. People's expectations of Bergen's climate – their identity as citizens of a city in which it 'always rains' – played an important role in defining the weather of a day or a season as good or bad.

## When Cultures Change

If climate is an idea that emerges from the relationship between weather and cultural ways of life, then as cultures change one might expect (imagined) climates to change. In other words, since notions of 'good' or 'bad', 'tolerable' or 'intolerable', 'safe' or 'dangerous' climates are bound up with distinctive ways of life, when social practices and cultural norms change, these imagined categories of climate might also change. Observed changes in climate–society relations, which superficially might be attributed to a change in physical climate, may on occasions be more accurately be attributable to changes in culture. For example, increasing vulnerability of the elderly to extreme heat might result from changes in care practices and population ageing rather than from more frequent or intense heatwaves. The growing prevalence of street cafés in temperate climates might result from changing eating habits or the affordability of patio-heaters rather than from a general warming of climate. Reductions in the number of

**Figure 5.1** Cartoons by Audon Hetland, an artist from Bergen, presenting the very special relationship of the local people with their weather. (top) 'Look, here goes the last Bergener'. (bottom) 'High tide at Bryggen' (Source: Meze-Hausken, 2007).
© Audon Hetland

abandoned professional football matches in winter might result from changing thermal infrastructure in sports stadia as much as from milder winters. I introduced a historical example of such a culturally induced change in climate in **Chapter 2**: Caribbean climates changed from being deadly and dangerous for eighteenth-century European settlers to becoming desirable and attractive for twentieth-century American tourists (see **Box 2.2**). Changes in culture alter people's interactions with the materiality of climate, which therefore changes people's imaginative engagement with the idea of climate.

A different example can be seen in changing cultural attitudes to climatic seasonality and the emergence of the medical condition known as seasonal affective disorder (SAD). SAD is today understood to be a biological response to seasonal changes of the hours of daylight. For some people it is generative of symptoms such as fatigue, weight-gain and depression. Rates of autumn–winter SAD appear to increase with latitude in the northern hemisphere, particularly in Europe and North America, suggesting that photoperiod length is a trigger. However, not all cultures appear to experience climatic seasonality in this way; neither has SAD been a condition that has affected European cultures historically, or at least not as a pathological condition. Rather, SAD seems to emerge in those cultures where 'natural time' (climatic seasonality linked to day-length) and 'social time' (culturally sanctioned seasonal rituals and practices) have grown apart (Harrison, 2004). Although not denying that changes in brain chemistry do likely occur, the argument here is that in some cultures these rhythmic changes in climate have not been, and are not, experienced as pathology. Indeed, climatic seasonality is attenuated, and hence normalised, through powerful symbolic associations between seasonality and meaning-making cultural or religious rituals.

How cultural change alters imaginative engagements with climate can also be seen in respect of new technologies of thermal comfort. This is most evident in terms of indoor climates where technologies of climate control have reduced the range of thermal variation to which many people are exposed and therefore come to expect as normal. Cultural variations here play a huge role in mediating norms and expectations of desirable indoor climates, with consequences for how outdoor climatic variation is perceived (witness the example of London's office workers cited earlier). But cultural change also asserts its influence through technological interventions in the liminal climates of outdoor private and public spaces, such as garden patios and street cafés. In many developed northern mid-latitude countries, recent years have seen the huge growth of the patio-heater as the idea of personalised outdoor climates has become normalised (Hitchings, 2007). Thermal control of indoor climates is extended to these outdoor spaces of casual socialisation, further adjusting cultural expectations of climatic normality, abnormality and control.

## When Climates Change

If ways of living with climate can change because cultures change, the corollary is that they also change because of changes in physical climate. People may experience a change in their weather during their lifetimes in two ways. People may move for periods of time to live, work or recreate in other regions with different climates. In this case, climate changes *ex situ*. Alternatively, the physical characteristics of weather in the places in which people remain living may change through their lifecourse. This is climatic change *in situ*. There is a long history of anxious reflection about how people change as a result of either of these experiences, whether bodily, emotionally, morally or socially.

European traders and settlers of the colonial era were particularly anxious about the effects of *ex situ* climatic change. Europeans had inherited from Classical thought the belief that climate could influence human physical, mental and moral health. If the temperate climates of Europe were conducive to temperate and morally desirable ways of living, how would northern Europeans fare as inhabitants of the torrid climates of tropical lands or as settlers in the Mediterranean climates of Australia and California? It was not so much the adjustments to material practices that were of concern, whether adapted building designs or new cultivation practices. It was rather worries about changes that hot and humid climates might engender in bodily health and moral hygiene. Influential in this regard was the English colonial surgeon James Johnson and his 1818 book *The Influence of Tropical Climates on European Constitutions*. Johnson argued that natural acclimatisation of European bodies to tropical climates would take generations, if not centuries. But his claim also was that changes in temperament and character associated with these bodily adaptations might be less than desirable. This narrative lent itself to a naturalised racial theory in which European intellectual and moral ascendancy over aboriginal populations was justified. Reflections on how people's ways of life and bodily health are altered by *ex situ* changes in climate brought about by human mobility continue in today's world, although less in the context of racial superiority. For example, German expatriates moving to Spain, Pakistani workers moving to the Gulf States or African students studying in Russia offer a new natural laboratory for exploring the physiological, psychological and cultural effects of weathering on the human subject.

More salient today it seems are concerns about potential changes to ways of life brought about by *in situ* changes in climate. The prospect of human-caused climate-change offers a scenario in which no land's physical climate will stay the same: climates will generally become hotter (or less cold) with associated changes in humidity, rainfall and snow. Cultural adjustments to *in situ* climatic change can be a response either to experienced changes in climate (e.g. changes in leisure or recreational pursuits as lifestyle adjustments to a changing physical environment) or else to presumptive climatic scenarios (e.g. an emotional sense of anticipated loss or psychological disturbance). The Australian academic Andrew Gorman-Murray expressed such sentiments after reading about scientific predictions for the end of the century for a radical reduction in snowfall and days of snow lying in the Australian Alps:

> I reacted 'emotionally' to the projections: I was shocked and disappointed. Especially about the imminence of climate-change impacts ... the emotional tenor of my response to projected snow loss was the catalyst for a research project... I was interested in the broader meanings and feelings associated with snow cover [in Australia]. (Gorman-Murray, 2010: 61).

Gorman-Murray went on to examine the reactions of Tasmanian residents to these putative changes in the snow climate of Australia. His study showed how the material, economic and emotional dimensions of the idea of climate were inseparable from the ways in which Tasmanians thought about their island. It was not just changes to livelihoods (snow tourism) or hydrology (earlier snow melt) that triggered anxiety, it was the imaginative attributes of Tasmanian climate which would also be reconfigured,

> When you live in Hobart, people talk about the mood of the mountain and snow is one of the important moods of the mountain ... it probably hits a lot of people in the face that that is no longer a mood which we're experiencing and we can't take it for granted any longer. (Gorman-Murray, 2010: 73–4)

## When Climates *and* Cultures Change

The two previous sections illustrate in different ways how human ways of living with climate adjust both to changes in culture and to changes in physical climate. Yet it is a central argument of this book that neither physical climates nor human cultures are static, both are in constant flux and mutually shape each other. It is therefore instructive to elaborate cases of adjustments to the idea of climate prompted by the simultaneous experience of climatic *and* cultural change. As people and ideas become increasingly mobile in a globalised world, it becomes routine to experience diverse climates and encounter different ways of living climatically, unlike the relatively rare experience of English expatriates in nineteenth-century India (see **Box 5.2**). Physical mobility arises from economic, educational or recreational motives (e.g. migrant labour, international study, global sporting events), while exposure to exotic climates and associated lifestyles can also result from vicarious encounters on digital platforms (e.g. global news channels and social media such as YouTube). Either way, there is today a form of climate cosmopolitanism probably unlike any other in human history.

---

**Box 5.2: An English Climate Far From Home**

British colonists in India in the eighteenth and nineteenth century were acutely aware of the differences between their new climates and those of their native Britain. Their everyday lived experience of the Indian tropical climate left traces on their bodies and mental dispositions and shaped certain forms of colonial social practice. But this experience of an exotic climate also interacted in complex ways with their imaginative lives, their morals, moods and sense of national and racial identity. These reflections can be

---

excavated through personal diaries and letters left by a variety of colonists, not least medical practitioners and governing elites.

The diaries of two such colonial figures in Bombay in the 1820s have been the subject of a detailed study by climate historian George Adamson. The personal papers of Mountstuart Elphinstone (the then Governor) and Lucretia West (the wife of the Chief Justice) revealed the importance of their lived experience of Bombay's monsoonal climate over a number of years. For Elphinstone, anxieties about the physical, mental and moral effects of this humid coastal climate – in his words, 'the languor of hot weather' – prompted him to found one of the early Hill Stations of the British colonists, at Khandala in the Western Ghats. At an elevation of 550 metres above the Indian plain, Elphinstone, West and other expatriates imagined themselves to be encountering a climate much more to their liking, more 'English' in its visual and sensual associations, than the 'enervating' climate of downtown Bombay.

In reality, however, this climate was far from 'English' in character. Yet discourses of tropical disease, nostalgia for England and the association between climatic and racial superiority together worked to construct a certain notion of a climatic Englishness in the minds of these diarists as they visited Khandala. How people live culturally with their climates can be a determined act of the imagination, as much as it may be a physiological response to atmospheric properties. British colonists such as Elphinstone and West resisted some forms of acclimatisation to India's 'degenerate' climate. But their attempt imaginatively to re-create an English climate far from home was deemed to be good for their physical health and moral vigour. It also contributed to shoring up their sense of racial superiority over native Indians. How one chooses to live with climate can be a political act.

(Sourced from Adamson, 2012)

In this cosmopolitan world one might suspect that cultural adaptations to climate and weather are themselves becoming more dynamic and varied. If the purpose of adapting to climate is to secure tolerable, satisfying and productive conditions for life, then a hyper-mobile world may be better placed than an insular one to secure these cultural goods in the face of climatic volatility. Yolande Strengers tested these ideas in a study of international students arriving in Melbourne, Australia, from 10 varied countries spanning cold, temperature and tropical climates. She wanted to see to what extent cultural ways of living with climate – clothing, diet, leisure, thermal control technologies – travelled intact with the students or whether these adaptive practices themselves adapted to new climatic and cultural surroundings. Studying the ways in which globally mobile students sought to stay warm or cool through the climatic year, these authors concluded that increasing exposure to varied weather conditions may well enhance rather than limit adaptation (Strengers and Maller, 2013). How people live with their climate is certainly learned in places and through cultures. And yet humans improvise and call upon a variety of material and immaterial repertoires – clothing, technologies, activities, memory, emotion, tolerance – in order to secure their personal adaptation goals. These might be as much about managing and tolerating interesting, exciting and challenging climatic conditions, than about more prosaic pursuits of thermal comfort and climatic safety.

## Chapter Summary

In this chapter I have shown that the idea of climate can fruitfully be understood as emerging from how people live materially and imaginatively with weather in particular places. Climate becomes a rich ensemble of atmospheric processes, material technologies, memories, landscapes, dress codes, social practices, symbolic rituals, emotions and identities. Taken collectively these climatic behaviours may be thought of as 'weather-ways'. These weather-ways embrace life in indoor and outdoor climates in specific places and also in the liminal spaces where new technologies extend the climatic imaginary. Indeed, how most people experience and understand their climate is through moving in and out of these multiple spaces. Place, memory, emotion and identity are woven into people's conception of climate. This lived reality of climate sits uncomfortably alongside more abstract and numerical depictions of climate emerging from scientific practice. In particular, framing climate and its changes as global in scale, the dominant frame in the discourse of climate-change, fails to connect with the more intuitive and culturally shaped ways of apprehending climate which this chapter has highlighted. The latter appears more faithful to the human experience of climate as something that can only fully be understood by paying careful attention to mind and place.

A corollary of understanding climate this way is that the human experience of climate never remains static. For example, many people today experience a change in climate as a result of hyper-mobility or through vicarious encounters with different climates mediated by digital platforms. Changes in cultural norms, social practices and new material technologies also alter the ways in which people come to understand and live with their climate. People experience changes in climate as they grow old simply because memories change, ways of life alter and their physical body ages. Yet physical climates – the systematic and rhythmic patterns of weather – also change over decades or more, and this too unsettles established notions of climatic normality. Changes in the weather provoke adjustments in 'weather-ways' not only because they are experienced physically, but also because they are imagined pre-emptively. Future climatic scenarios have the potential to unsettle these embodied and imaginative engagements with climate, something I explore further in **Chapter 7**. But before then I turn in **Chapter 6** to the question of blame: what range of things have been, and still are, blamed on climate?

## Further Reading

Janković, V. and Barboza, C. (eds) (2009) *Weather, Local Knowledge and Everyday Life: Issues in Integrated Climate Studies*. Rio de Janeiro: MAST.

Mabey, R. (2013) *Turned Out Nice Again: Living with the Weather*. London: Profile Books.

Meyer, W.B. (2000) *Americans and Their Weather*. Oxford: Oxford University Press.

Sherratt, T., Griffiths, T. and Robin, L. (eds) (2005) *A Change in the Weather: Climate and Culture in Australia*. Canberra: National Museum of Australia Press.

# 6

# Blaming Climate

## Introduction

One of the questions that the idea of climate provokes is the extent and nature of its role as a causal agent in the world. What physical and social phenomena are caused by climate and its changes? What sort of things can be 'blamed' on climate? And the enduring question 'Why is "the Other" different?' – to which the seductive explanatory power of climate to explain human difference has frequently been invoked. Sages, philosophers and public intellectuals have long deliberated these questions and the cultures into which they speak have frequently been influenced by the answers given. But the range of phenomena which are said to be caused by climate – or as understood more recently to be caused by climate-change – extend well beyond questions of racial or cultural 'othering'. Why are peatlands deteriorating? Why are some economies more productive than others? Why does coral bleaching occur? Why are people migrating? The list of outcomes for which climate and its changes has been called upon to take responsibility, or in which climate is 'implicated', seems near endless. I start this chapter with one example of where this search for climatic blame can lead.

In 1867, the president of the medical college of New York University, John William Draper, published his *History of the American Civil War* in three volumes (Draper, 1867). Draper was a determinist, believing that societies are guided by 'uncontrollable causes'. Chief amongst these causes was climate, a subject which preoccupied the first volume of his *magnum opus* in which he laid out the causes of the Civil War,

> The nations of men are arranged by climate on the surface of the earth in bands that have a most important physiological relation. In the torrid zone, intellectual development does not advance beyond the stage of childhood... In the warmer portions of the temperate zone, the stage of youth and commencing manhood is reached... Along the cooler portions of that zone, the character attained is that of individual maturity, staid sobriety of demeanor, reflective habits, tardy action. (Draper, 1867 (vol. 1): 101–2)

With such a view of the world it was easy for Draper to see the hand of climate at work in laying down the conditions which had provoked the bloody Civil War between 1861 and 1865 and which had led to the deaths of 600,000 Americans. Climate had separated America into two nations, into two political economies, into two moral communities. By *favouring* plantation life in the American south, climate naturalised the institution of slavery. By *denying* the possibility of plantation life in the north, climate explained the Unionist political challenge to slavery. 'Climate tendencies facilitate the abolition of slavery in a cold country, but oppose it in one that is warm' (1867 (vol. 1): 342).

David Livingstone's account of Draper's deterministic ideology emphasises that he was far from alone in thinking this way (Livingstone, 2015). Philosophers before Draper, for example, Montesquieu, and geographers after him, for example, Ellen Semple, held similarly robust views about what climate could and could not explain. The consequence of blaming the American Civil War on climate was to deflect attention away from human judgement and political choice. As summarised by Livingstone, 'Climate's actions had the effect of freeing political history from the burdens of moral accountability' (2015: 439). The Civil War was not simply *caused by* climate; it could be *blamed on* climate, a narrative which released morally culpable actors from their role in triggering the conflict. History was reduced to geography, politics to fatalism, morality to nature.

* * *

This chapter is concerned with the question 'For what things can climate be blamed?' Although straightforward to ask, this question has attracted a multiplicity of answers through the ages, many of them unhelpfully simplistic. There is a long history of elevating climate as the (primary) determinant of human physiology and psychology, as much as climate has been elevated as being the chief determinant of physical landscape and biological evolution. Climate has been 'blamed' for wars, economic growth, street violence, political despots, famine, property prices, suicides, the age of menstruation – and many more phenomena. Although climate determinism is easy to caricature and easy to dismiss as hopelessly naïve, there are more sophisticated variants circulating today which elsewhere I have referred to as 'climate reductionism' (Hulme, 2011). Climate reductionism offers a methodology for providing simple answers to complex questions about the relationship between climate, society and the future. In its crudest form it asserts that if social change is *unpredictable* and climatic change *predictable*, then the future can be made known by elevating (predicted) climate as the primary driver of physical and social change.

Discerning who and what has agency in this tangled matrix of socio-natural life requires deep understanding of cultural norms and practices; how the human mind and body, and how social institutions, engage their external worlds. The chapter explains how different cultures have understood this difficult problem of climatic agency, first by examining theories of climatic causation and then by giving examples of things for which climate, and climate-change, has been blamed.

## Climatic Cause, Blame and Culpability

In **Chapter 2** I emphasised the importance of the distinction between understanding climate as a descriptive index (what climate *is*) and seeing climate as a causal agent (what climate *does*). Here I want to explore more carefully what it means to say that climate (or climate-change) causes something to happen. What type of agency does climate have? And where, if at all, does moral culpability lie?

Causal reasoning is essential to scientific inquiry. To understand the physiological causes of a human disease for example, the body is reduced to smaller and smaller units to enable controlled experimentation and thereby to find connected chains of cause and effect. An excellent example of such a goal is the European Union's Human Brain Project, with its ambition 'to establish *in silico* experimentation as a foundational methodology for understanding the brain'. A similar scientific search for cause and effect is underway with regard to the Earth's climate system. Computational climate models seek to simulate the world of material cause and effect by representing all significant physical processes in mathematical form. But causal reasoning is also essential in social life, in domains such as legal reasoning and public policy intervention. Did the driver's alcohol intake cause the fatal car accident? Will placing a tax on carbon emissions cause a reduction in the combustion of fossil fuels? In most social situations, however, there will rarely be *one* single cause of a particular outcome. In the case of the car accident not only the driver's alcohol level, but the condition of the car, the state of the weather, the signage of the road and the behaviour of the pedestrian all need to be taken into account. In most social situations multiple factors are at work and single causes cannot be isolated through controlled experimentation. Each event is unique, occurring within its own evolving contextual web.

The above illustrations help isolate the difference between cause, blame and culpability, terms relevant in reflecting on the morality of acting in the world. To say $x$ causes $y$ is to be neutral with regard to the desirability or otherwise of outcome $y$. That carbon dioxide in the air causes heat to be trapped in the atmosphere is neither here nor there. It is mere fact. But the language of blame implies the passing of some judgement on outcome $y$. To say that the increasing heat of the atmosphere is *blamed on* carbon dioxide is not merely a statement of cause and effect. It is that, but it is more. It is to say that this outcome – a warmer atmosphere – is in some way, and for some interested party, a deleterious outcome. Further to this comes the question or moral culpability, of blameworthiness. If the excess in carbon dioxide in the air is the result of a volcanic explosion one would not claim that the volcano is morally culpable. But the situation is different if the carbon dioxide arrives in the air because of wilful combustion of fossil fuels by a human agent, in full light of its injurious consequences. Here, the combustion of fossil fuels is the action 'to blame' for the excess of carbon dioxide, but one might be entitled to claim also that the human agent that perpetrates the action is morally culpable for the consequences.

There is a powerful human desire to find causal explanations for destructive outcomes, in other words to construct narratives of blame. Even more is there a desire to isolate blameworthy agents, to identify morally culpable actors who can be held accountable for these outcomes. Such narratives are societies' means of coming to terms with disaster and catastrophe. Without culturally credible and unifying accounts of blame the world would be too terrifying and capricious a place in which to live sanely. Thus it is helpful for a society to blame the flooding on heavy rain. More satisfying however, at least in some cultures, might be to blame the flooding on an action of God or on sinful human behaviour which provoked God into action (see **Chapter 4**). More satisfying still perhaps, and certainly so for contemporary secular cultures, is to blame the flooding on incompetent drain maintenance, poor land use planning or inadequate flood defences. In modernist cultures such as these, '... [the] target [for blame] has moved from indifferent nature and sinning victims, to criminally negligent magistrates or persistent social inequities' (Janković, 2006: 40). Indeed, such reasoning is one of the hallmarks of Ulrich Beck's risk society (see **Chapter 7**). Nothing can be accepted as a mere accident or an 'Act of God'. Increasingly, there is a need for greater and greater accountability for all adverse physical and social phenomena, and in many cultures this has been accompanied by the growth of litigation. Social and political institutions seeking to control individuals need to identify responsible agents who can be rewarded or punished accordingly.

This brief excursion into theories of blame provides a context for the two sections to follow. First, I elaborate situations, both historical and contemporary, where climate has served as a useful causal explanation for particular conditions or outcomes. Second, I consider cases where the idea of climate-change has satisfied the need, not just for casual explanation, but for identifying agents who can be blamed for deleterious outcomes. This achievement introduces the possibility, at least theoretically, of accountability, redress and compensation for damage caused. My examples explain why such climatically rooted causal or blameworthy explanations of phenomena in the world are pervasive, persistent and, in some sense, comforting.

There is one further aspect of climate as a causative agent to which I would like to draw attention. This is an argument which has recently been made by the historian Vladimir Janković and concerns the idea of fetishism. According to Simpson (1982: xiii), fetishism occurs when the mind '.. ceases to realise that it has itself created the outward images or things to which subsequently it posits itself as in some sort of subservient relation'. It is therefore possible, so Janković (2014) argues, to think of climate as a fetish: an idea invented by humans – which is a core argument of this book – and one in which we imagine ourselves to be in a submissive relationship. As a fetish, climate is granted an unwarranted power over the human imagination, bodily physiology, material artefacts and social institutions. It is this oppressive relationship between people and climate which I explore in the next two sections.

## Blaming Climate

There is a long tradition of western thinkers valorising certain climates as being conducive to virtuous character and the higher forms of civilised life (see **Chapter 2**). This has been a dominant trope when thinking humanly about the relations between climate and people and is summarised, perhaps facetiously, by the historian Clarence Glacken:

> ... warm climates produce passionate natures; cold [climates], bodily strength and endurance; temperate climates, intellectual superiority; and among the non-physiological theories, a fertile soil produces soft people, a barren soil makes one brave. (Glacken, 1967: 81)

Yet this enduring trope is less about blame for an adverse outcome than it is about a justificatory explanation for desirable human attributes. Certain people, certain regions, certain societies are innately superior to others because of the gift of the climates in which they dwell. Climate becomes a means of grace rather than a source of blame. Although many such arguments have been couched as description of mere fact, too easily and too often they have been used as justification for human arrogance, domination and exploitation.

A different frame for approaching climate's causative powers is that of cultural introspection. Depending on one's point of view, climate is either the giver or confounder of bounty and good fortune. Contrast these two examples of cultural reflection from British daily newspapers in the mid-nineteenth century:

> The climate of Britain [...] perfects all the substantial necessaries of life required from the soil; and it has given an athletic frame, and impressed an energy and perseverance of character on the inhabitants, which never could have been developed amid the lassitude of an oriental climate, or beneath the rigour of the Northern sky. (*Daily News*, 1850)

> Our variable climate has asserted its power of counteracting the labours and defeating the hopes of our people, and threatens to snatch from us the fruit of so much industry and skill just when we were about to gather it... To trust the subsistence of so populous a nation to the mercy of a climate so fickle and so variable as ours [...] would be an act of desperate follow and suicidal wickedness. (*The Times*, 'The unseasonable weather of the last few days', 18 August 1852)

Either way, whether Britain is ennobled or weakened by climate, these public commentators are bowing to climate as fetish, elevating its mystical powers to hold sway over the moral and economic fortunes of the nation.

Climate has also frequently been invoked to (help) explain the decline and/or collapse of civilisations. One of the first notable invocations of such historical climatism was Edward Gibbon's account of the decline and fall of the Roman Empire. Gibbon hinted that the declining agricultural yields which weakened the agrarian economy of the Empire had a climatic origin, as did the migration of Huns and Goths from central Asia in the fourth century CE which resulted in military confrontation with the Romans. More recent examples of this trope have focused on the Mycenaean civilisation in Greece, the First Nation Indians of the American Great Plains and the Malian Kingdom in West Africa. In their study *Climates of Hunger*, Reid Bryson and Thomas Murray offered detailed case studies of how sustained episodes of drought had undermined the stability of these three societies (Bryson and Murray, 1977). Similarly detailed archaeological and palaeoclimatic work has suggested that the demise of the Mayan civilisation around the ninth century CE was triggered by a severe drought. There are interesting parallels here between the inevitable fates of societies subject to these climatic forces and those succumbing to the unfolding of Marx's historical materialism (according to which the material conditions of a society's economic mode of production determine its future development).

Apart from explaining racial difference (i.e. superiority) or societal collapse, a third deployment of climate as cause comes in the service of masking or absolving political accountability. One instance in which climate was made to do political work was the League of Nations' debate in the inter-war years about human trafficking (see **Box 6.1**). But there are many other examples, such as accounting for deaths due to desertification in Sudan, famine in India or heatwaves in American cities. It may be politically 'tidy' and convenient to use climate in naturalistic explanations for such social disasters, but it has an eviscerating effect on political accountability. In each of these illustrative cases, powerful and persuasive counter-narratives to the 'climate-as-blame' account have been offered: respectively, Amyarta Sen's *Poverty and Famines: An Essay on Entitlement and Deprivation* (Sen, 1981), Mike Davis's *Late Victorian Holocausts: El Niño Famines and the Making of the Third World* (Davis, 2002) and Eric Klinenberg's *Heatwave: A Social Autopsy of Disaster in Chicago* (Klinenberg, 2002). The buck-passing is neatly captured in Anna Carlsson's study of London flooding in 1928 which led to the death of 14 people and widespread damage. This became a topic of considerable public debate in the weeks following the flooding, the local authorities blaming exceptional weather conditions rather than any lack of preparedness planning. A national newspaper, *The Daily Mirror*, was quick to respond:

> According to themselves, the authorities are to be excused – because 'no record exists' of conditions like these. Despite a common British delusion that our climate [...] is, on the whole, equable, mild and, as

we might say, "normal" [...] we *seem* always to be grappling amateur-
ishly with weather alleged to be *exceptional*.' (Quoted in Carlsson,
2009: 92; emphasis in the original)

The outcome of the various official inquiries into the events was that climate,
and not London's planners, was to blame, 'a consequence of natural or divine
forces that could not be defended against' (Carlsson, 2009: 94).

---

## Box 6.1: Climate and the Age of Consent

In the summer of 1921 the first League of Nations Convention on Trafficking
met in Geneva, attended by delegates from 34 states. Their aim was to
ratify the 1910 International Convention for the Suppression of White Slave
Traffic. The underlying ambition was to protect girls from trafficking for prostitution and to do so by establishing a universal age below which practising
prostitution would be illegal. Establishing an international norm for the age
of sexual consent at 21 years would allow more traffickers to be prosecuted.
But the meetings quickly ran into trouble regarding the very different cultural
norms about the ages of consent and legalised marriage which prevailed
in different nations. A proposal to exempt 'Eastern countries' and 'tropical
colonies' from the age standard was rejected. No agreement was reached
in Geneva in 1921 and negotiations were to continue at a series of further
meetings through the 1920s and early 1930s.

These on-going discussions about differences in the age of sexual
activity focused around the age of first menstruation, or the menarche.
And it was here that climate, in particular the average temperature of
nations, entered the discussions as a useful explanatory force. French,
Polish and Italian delegates used climate as a short-hand to capture differences between the sexual mores of various nations, specifically the
menarche. They successfully argued against a single international standard and instead lobbied to have differences between nations rest on the
explanatory variable of climate. For example, at a 1930 meeting the Italian
delegate noted that it was 'unjust to fix an international age for minors,
since social and climatic conditions tended to alter this age in different
countries' (quoted in Tambe, 2011: 114). While it was clear to all par-
ties that countries differed in their sexual practices, invoking the idea of
climate as controlling the menarche lent a naturalised certitude to these
defensive justifications. In the end, this assertion and acceptance of
climatically determined differences in the menarche inhibited the League
from reaching a consensus on an international age of consent, thwarting
one of the goals of the anti-traffickers.

Climate served as a gentle means to express the competing imperialist
nationalisms that were at work within the League. In this arena, where dip-
lomats sought to arrive at a consensus, naturalised explanations of the age
of consent rendered cultural differences more palatable and less disputable.

> Climate offered a convenient index of national differences in sexual practices, drawing upon the sciences of race which the nineteenth century had put into circulation. References to climate's imputed power performed important ideological work in naturalising hierarchical relations between nations. Climate, for neither the first nor last time, became enlisted in biopolitics – those strategies and means by which regimes of authority seek to regulate human life processes.
>
> (Sourced from Tambe, 2011)

## Blaming Climate-Change

While climate has long been an explanatory category in cultural discourses of disaster causation and blame, climate has not so easily been offered up as a blameworthy *moral* agent of social disaster or disorder. Or at least when it has, climate has merely been the proxy for the culpable agent in the background: usually God, the gods or the spirits, whether or not provoked by human behaviour. But the more recent idea of climate-change, by which I mean changes in physical climates caused naturalistically by human modification of the atmosphere (see route 4 in **Figure 4.2**, p. 47), offers a different twist. The culpable agent for climate-change is to be found not in the spiritual realm but in the world of humans, whether this be all humans, rich humans, westernised humans, the capitalist system or technological hubris. Invoking climate-change as a cause of adverse social or physical phenomena offers the prospect that culpable agents can be held to account. In this sense, climate-change becomes an attractive naturalistic explanation for the ills of the world since it satisfies the human need for accountability. It is neither the gods nor spirits, nor the even less accountable 'forces of nature', that are to be judged, but in a very modernist way it is one set of people who can hold another set of people to account (see **Chapter 4**). Given the sheer scale of many of the pervasive social, ecological and economic ills that assail the world today, it is the equivalent scale of climate-change (i.e. something which operates globally) which explains the astonishing success of this narrative of blame. In this sense climate-change takes on, ironically, almost god-like powers. If climate-change is not a fetish, it is in danger of becoming like one.

The ubiquity of climate-change as blameworthy is easily illustrated by following any number of websites[1] and media platforms today (see **Figure 6.1**). Out of the many hundreds of examples I could have chosen, I select here just six which sample the range of attributed phenomena which have either already

---

[1] For an example see this website: 'A complete list of things caused by global warming', http://www.numberwatch.co.uk/warmlist.htm (accessed 12 May 2016).

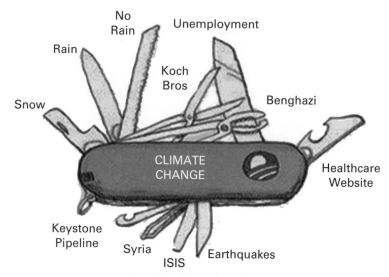

**THE EXCUSE FOR EVERYTHING**

**Figure 6.1** Climate change: the excuse for everything (Source: Watt's Up With That, September 2014 – https://wattsupwiththat.com/2014/09/30/climate-change-created-isis-is-now-49-on-the-official-list-of-things-supposedly-caused-by-global-warming/ (accessed 19 May 2016)).

occurred or else prospectively are claimed will occur because of climate-change (or its synonym global warming). This small sample of examples has been selected not necessarily for its scientific rigour nor on the basis of the evidential standards of validity being used. What I am interested in illustrating is the cultural, psychological and political work that is being achieved by these public claims of climate-change blameworthiness.

- *'Global warming* is helping America's sworn enemies Al Qaeda as well as the Taliban in Afghanistan'[2].

- *'Climate change* has brought once lively and loud habitats to utter silence as their inhabitants of birds, frogs and insects have either vanished or drastically changed their migration patterns'[3].

- *'Climate change* is driving species shifts both on land and on sea' (Blanchard, 2015).

---

[2] ABC News, 10 October 2009: 'Al Qaeda and Taliban being helped by global warming', http://www.newsbusters.org/blogs/noel-sheppard/2009/10/10/abc-al-qaeda-taliban-being-helped-global-warming (accessed 12 May 2016).

[3] AccuWeather, 19 October 2013: 'How is climate change jeopardising the sounds of nature?' http://www.accuweather.com/en/weather-news/climate-change-jeopardizes-the/18852750 (accessed 12 May 2016).

- 'There is clear indication that as the tensions of *global warming* continue to heat up so may the possibilities of war... a Hot War! ... a World War IV'[4]. See also **Box 6.2** for another example of the 'climate-change causes war' narrative.

- *Global warming* may make the world's inhabitants cranky and stressed ... give them cancer and even worsen their suffering from sexual dysfunction'[5].

- By 2030, *climate change* will increase the risk of conflict in Africa by 54 per cent which 'would result in a cumulative additional 393,000 battle deaths' (Burke et al., 2009).

---

## Box 6.2: Climate-Change and the Syrian Civil War

In the last few months of 2015, in the weeks before the COP21 climate negotiations in Paris, a large number of social commentators, politicians, NGOs and government commissioned reports began to openly associate the civil war in Syria with climate-change. One notable voice was that of Prince Charles, the heir to the British throne. He told Sky News, 'There is very good evidence indeed that one of the major reasons for this horror in Syria was a drought that lasted for about five or six years', adding that climate-change is having a 'huge impact' on conflict and terrorism[6]. Others too, President Obama, Al Gore and democratic Presidential candidate Bernie Sanders, also talked of a link between climate-change and the Syria conflict, Sanders going so far as to argue that 'climate-change is directly related to the growth of terrorism'[7].

This is not the first time that war has been blamed on climate-change. In 2007, the General-Secretary of the United Nations Ban Ki Moon contended that 'the Darfur conflict began as an ecological crisis, arising at least in part from climate-change'[8]. The argument behind such claims runs like this:

*(Continued)*

---

[4] Treehugger, 4 January 2009: 'Will global warming cause World War IV?' http://www.treehugger.com/corporate-responsibility/will-global-warming-cause-world-war-iv.html (accessed 11 December 2015).

[5] WND News, 5 January 2010: 'Not hot! Your sex life hit by global warming', http://www.wnd.com/2010/05/147617/ (accessed 12 May 2016).

[6] Mills, R. (2015) 'Charles: Syria's war linked to climate change', Sky News (23 November); available at: http://news.sky.com/story/1592373/charles-syrias-war-linked-to-climate-change (accessed 11 June 2016).

[7] Schulman, J. (2015) 'Bernie Sanders: yes, climate change is still our biggest national security threat', Mother Jones (14 November); available at: http://www.motherjones.com/environment/2015/11/bernie-sanders-climate-change-isis (accessed 11 June 2016).

[8] Ki Moon, B. (2007) 'A climate culprit in Darfur', *The Washington Post* (16 June); available at: http://www.washingtonpost.com/wp-dyn/content/article/2007/06/15/AR2007061501857.html (accessed 11 June 2016).

*(Continued)*

human emissions into the atmosphere are changing the world's weather patterns; droughts in dryland regions are therefore becoming more frequent or more severe; this reduces access to scarce resources such as water or food; people are consequently forced to leave their land moving either to cities (Syria) or into the territory of different ethnic groups (Sudan); civic tension, conflict and war is the result. The reasoning is then extrapolated to claim that such mass migration and civil conflict is only a foretaste of what is to come because of future climate-change. For example, Lord Ashdown, the leader of Britain's Liberal Democrats for 11 years and a former army officer, speaks thus: 'The numbers we now have of refugees fleeing battle zones are going to be diminished into almost nothing when we see the mass movement of populations caused by global warming'[9].

   I cite this example of blaming climate-change, not because the evidential basis for the claim is necessarily strong. In fact it isn't, as a host of academic studies have shown in relation to the Darfur war. Claims that 'climate-change causes war' need to be tempered with thicker cultural and political analyses of how social relations become stabilised and destabilised. Rather, I include the example because it is a particularly egregious and culturally powerful example of public claims connecting climate-change with war. As I showed earlier in my examples of the American Civil War and the League of Nations (see **Box 6.1**), climate (or in this case climate-change) is mobilised to undertake political work. If mass migration, conflict and war are believed to be caused by climate-change rather than by political oppression, ethnic conflict or corrupt resource management, political and diplomatic interventions are redirected in ways which serve some interests, but not others.

These examples serve the human need for causal judgements about adverse outcomes which serve two primary functions: a backward-looking function which assigns blame for harmful events in the past and a forward-looking function which seeks to avoid harmful events in the future. Climate-change is a powerful force shaping the world precisely because it seems to offer a unitary account of blame. What is causing the Taliban to thrive, birds to fall silent, fish species to move, Syrians to go to war, humans to lose sexual functionality, Africans to kill each other? It is the force-field of climate-change that connects these contemporary ills into a unifying account of blame and accountability. It offers the prospect that if climate-change can be stopped then the world would become a safer, more just and desirable place to live. This is a meta-narrative of mythical dimensions, hence its widespread cultural appeal.

---

[9] Truthdig, 9 September 2015: 'World must avert devastating flood of climate refugees', http://www.truthdig.com/report/item/world_must_avert_devastating_flood_of_climate_refugees_20150909 (accessed 12 May 2016).

This narrative works in similar ways to shift political responsibilities in ways similar to how 'climate' was blamed for flooding in London, in the example cited earlier. The work of Indian cultural anthropologist Nayanika Mathur is illustrative of the general case (Mathur, 2015). In the Himalayan borderlands of northern India, human–animal conflicts have long been a cause of concern. Recent disputes between villagers and state officials have erupted because of the increase in incidents of leopard killings of children and adults, increasingly aggressive Himalayan bears and dwindling numbers of musk deer, a protected species. Mathur is able to show how climate-change was used as a category of explanation which served political interests at state and federal level. The aggression of leopards and bears, and the vanishing musk deer, were all blamed on climate-change. For the man-eating cats and aggressive bears, climate-change could deflect attention away from state officials' responsibility to provide trained hunters and adequate equipment for managing wildlife. In the case of the musk deer, climate-change provided convenient cover for state-sanctioned poaching. Climate-change was covering up various sorts of misdoings of the local state. As Mathur explains:

> Agents of the state were actively trying to harness the persuasive power of climate change to produce a language that would be understood and accepted by their superiors sitting in the distant state capital... There is something peculiarly persuasive about climate change [as a form of explanation], especially when solemnly pronounced by formalised institutions such as states or international bodies. (Mathur, 2015: 101)

## Chapter Summary

There is a long history of elevating climate as the (primary) determinant of human physiology and psychology. There is also a history in which climate is offered as the primary determinant of physical landscapes, biological evolution, human conflict and economic prowess. Culturally credible and persuasive accounts of blame and culpability fulfil an important human social and psychological need. Such collective sense-making in a complex and chaotic world enables social institutions to function and societies to be governed. In this chapter I have examined some of the ways in which diverse social and physical phenomena are blamed on climate and its changes. This is nothing new. The range of things which historically have been caused by, or blamed on, climate is extensive. Wars, economic performance, street violence, political despots, famine, property prices, suicides, the age of menstruation – and many more phenomena – have all been 'explained' by climate. I have also explored and sought to explain why the contemporary phenomenon of climate-change has become such a pervasive and attractive category of blame in today's world. Yet given the intractability of complex social phenomena, sometimes referred to as 'wicked problems', there are dangers in elevating climate in explanatory accounts of the state of the world (climate determinism) or indeed in reducing the future to climate (climate reductionism). In either case, some manifestations of human agency and some forms of political accountability are downgraded, leading to partial and unstable accounts of social, economic and environmental realities. In the following chapter I turn my attention from examining ways in which climate is brought into narratives of blame, to how climate is invoked in narratives of fear, terror and risk.

## Further Reading

Clark, N. (2011) *Inhuman Nature: Sociable Life on a Dynamic Planet.* London: Sage.

Fleming, J.R. and Janković, V. (eds) (2011) 'Klima', *Osiris* 26 (1).

Rudiak-Gould, P. (2015) 'The social life of blame in the Anthropocene Environment and Society', *Advances in Research* 6: 48-65.

# 7

# Fearing Climate

## Introduction

On Saturday 3 February 2007, the day after the United Nations' IPCC published its Fourth Assessment Report, one of the UK's five quality national daily newspapers ran a full front page story warning of the consequences of a changing climate. *The Independent* offered five different climatic scenarios for the year 2100, five worlds warming in one degree increments by between 2.4 and 6.4°C. Printed against a backdrop of an orange-coloured fireball Earth, they pronounced what the high-end of this range would mean for life on Earth:

---

*Final warning. According to yesterday's UN report, the world will be a much hotter place by 2100. This will be the impact ...*

**[of 6.4°C warming] Most of life is exterminated.**

Warming seas lead to the possible release of methane hydrates trapped in sub-oceanic sediments; methane fireballs tear across the sky causing further warming. The oceans lose their oxygen and turn stagnant, releasing poisonous hydrogen sulphide gas and destroying the ozone layer. Deserts expand almost to the Arctic. 'Hypercanes' (hurricanes of unimaginable ferocity) circumnavigate the globe, causing flash floods which strip the land of soil. Humanity reduced to a few survivors eking out a living in polar refuges. Most of life on Earth has been snuffed out, as temperatures rise higher than for hundreds of millions of years.

(Sourced from: Lynas, 2007)

---

This was a portrait of a cataclysmic climatic future, resonant with depictions of the end of the Earth and the Second Coming of Christ found in the final book of the Christian Bible. In the Book of Revelation, a.k.a. The Apocalypse of John, one finds descriptions such as: 'hail and fire, mixed with blood, came pouring down on the earth'; 'a third of the living creatures in the sea died'; 'a third of the water turned bitter and many died from drinking it'; 'the sunlight and the air were darkened by the smoke from the abyss'[1]. Such climatic excesses, real or imagined, and the fearful reactions they provoke have punctuated human history. The terror experienced by people in southern England, as a violent storm ravaged its way across the land in July 1783 following the eruption of Icelandic volcano Laki, is but one example: '...the women shrieking and crying, were running to hide themselves, the common fellows fell down on their knees to pray, and the whole town was in the utmost fright and consternation' (*Exeter Flying Post*, 10 July, 1783; cited in Grattan and Brayshay, 1995: 130). It is not clear how the editors at *The Independent* extracted their headlines from the rather more sober report of Working Group I of the IPCC published the previous day, but the purpose of such an exaggerated interpretation of climate science is surely a familiar and well-worn one: it is the trope of environmental apocalypse (Buell, 2004).

Human cultures have always been capable of constructing narratives of fear around direct or vicarious experiences of strange, dangerous or portended climates. From early Greek notions of the uninhabitable torrid climates of the equatorial zone or the portentous climatic events of Shakespearian England (see **Box 2.1**), to James Lovelock's 2006 book *The Revenge of Gaia*, disorderly climates have the potential to unsettle the mind and induce anxiety, fear and terror in human populations. Historian Lucien Boia has described the instinct thus: 'The history of humanity is characterised by an endemic anxiety ... it is as if something or someone is remorselessly trying to sabotage the world's driving force – and particularly its climate' (Boia, 2005: 149). The language of the contemporary discourse about climate-change, illustrated above by *The Independent*, routinely uses a repertoire which includes 'danger', 'tipping points', 'collapse', 'catastrophe', 'terror' and 'extinction'. Such a breakdown of the world's climate becomes easily associated with the four horsemen of The Apocalypse of John: war, famine, pestilence and death.

\* \* \*

In this chapter I examine discourses of fear associated with climatic disorder extracted from different historical eras. Such discourses are always situated historically, geographically and culturally. These fearful climatic imaginaries are a response of the mind, shaped by particular cultures, to experienced or projected natural events. They are not unavoidable imprints of nature *on the*

---

[1] From the Biblical Book of Revelation; selected from Chapter 8, verses 7 to 11 and Chapter 9, verse 2.

human mind. Discourses of fear prompted by disorderly climates are an authoritative claim to power, as well as being an appeal to action, whether personal or political transformation (repentance or revolution). Understanding the cultural dimensions of discourses of fearful climates also offers insights into the impulse for purposefully redesigned climates (see **Chapter 10**) and how climatic futures are imagined (see **Chapter 12**). Yet however contemporary climatic fears have emerged – whether through resistance to neo-liberal globalism, through the emergence of risk society or through instinctive human anxieties about the future – it is through changes in culture that they will be assimilated, reconfigured or dissipated.

The chapter is organised in three sections. First, I examine how changes in climate are associated with narratives of existential threat; second, I explore the relationship between climatic volatility and extremity and idea of sublime terror; and, finally, I consider how in the present moment climatic anxieties exemplify Ulrich Beck's notion of risk society.

## Existential Threat

The archetypical account of climatic disaster and existential threat, at least for middle-eastern and western cultures, is the flood myth. I have already examined this in **Chapter 4** in the context of causative accounts of climatic change. The biblical (or Noahian) Flood, written in the second millennium BCE, was caused by unprecedented and divinely ordained rainfall. It led to the destruction of all sentient life, barring Noah, his family and breeding pairs of living creatures saved through God's mercy. A similar outcome occurs in the Sumerian flood myth from the fourth millennium BCE, humanity only being rescued by the boat-building exploits of the philosopher-king Atrahasis (Salvador and Norton, 2011).

Climatic change as existential threat continues to reside in contemporary cultural imaginations and the global flood remains one of the most widely known mythic narratives capable of realising such threat. It is evoked in many popular fictional writings and film scripts, such as J.G. Ballard's novel *The Drowned World* (1962) and recent movies such as *The Day After Tomorrow* (TDAT) and *The Age of Stupid* (see **Chapter 8**). The Hollywood blockbuster disaster movie TDAT, released in 2004, uses the flood myth to dramatise an apocalyptic account of prospective climate change (see **Box 7.1**), whilst the religious NGO Operation Noah explicitly draws upon the Noahian flood for its identity and faith-inspired campaigning mission. The influential environmental scientist James Lovelock also alluded to the remnant of the flood – 'the few breeding pairs' – when publicising his book *The Revenge of Gaia*: 'Billions of us will die [from climate change]', he claimed 'and the few breeding pairs of people that survive will be in the Arctic where the climate remains tolerable by the end of the twenty-first century' (Lovelock, 2006).

---

## Box 7.1: Climate and Apocalypse

Environmental movements have frequently drawn upon apocalyptic imaginary to induce fear and to alert people to dangerous realities or to pending future catastrophes. Frederick Buell's book *From Apocalypse to Way of Life* (Buell, 2004) is useful in tracing the history of such rhetoric and how it has appealed to people on both the political left and right. Yet the idea of apocalypse in the western imagination has a much richer set of meanings than simply a dramatic and convulsive ending to a physical world and the fear and loss which this entails. Buell identifies four features of the Jewish-Christian apocalyptic tradition which together help explain the richness of using the idea of apocalypse to transform fearful and fatalistic thinking about climatic disorder. First for Buell is the element of a sudden rupture or dislocation with the past. Second, a sober narrative of how this rupture will lead to an ending, often violent and disorienting, if no change is enacted. Third, a cosmic revelation – gnostic and mystical – of the true meaning of this rupture and ending. This revelation contains within it the hope and promise of redemption, a sense that the future *can* be changed. And then, fourth, is the idea of a final judgement, of a separation by the omniscient creator of the saved from the damned. When applying such a theological reading of apocalypse to historical and contemporary fears about changing climates, one can see some similarities in narrative structure, but also some opportunities for reinterpretation of what is at stake.

Stefan Skrimshire (2014) suggests three directions in which these theological insights into the apocalyptic tradition might be applied to the case of climate-change. First is the recognition in apocalyptic narratives of a deep pessimism towards human structures and powers, the realisation that climatic disordering challenges the modernist illusion of human control. Second, and perhaps paradoxically, is that apocalypse emphasises the dramatic role to be played by humans, for people to identify what constitutes an appropriate moral response to such events and to choose to be on the side of good in an unfolding cosmic drama. The moral response to an apocalypse varies according to cultural analyses of climatic causation. After the fear brought about in England after the terrible storm of November 1703, the national response was a day of repentance before God, led by the King and other dignitaries. Today, it is more likely to be a call for social justice, for restraint on conspicuous consumption or for cultural transformation. Third, apocalypse can also be cathartic, a disclosure of what *really* is wrong with the world. '[Apocalyptic] revelation … means knowing the truth about a corruption in social relations, in relations with nonhuman life and the false idol of world domination' (Skrimshire, 2014: 244). To interpret climatic disruption as an apocalypse is not merely to be fearful or fatalistic; it is a call to pay attention to what needs changing, in oneself and in one's relations with the world.

---

A different fear narrative originating in the ancient world involved the idea of uninhabitable climates, induced through either excess or deficient heat. Aristotle, for example, divided the world into five climatic zones, symmetrically distributed on each side of the equator. Two of these zones were called *oikou-menai*, the habitable world: the temperate zones in each hemisphere lying between the Tropics and, respectively, the Arctic and Antarctic circles. Straddling the equator was the equatorial zone demarcated by the Tropics of Cancer and Capricorn. According to Aristotle, this torrid zone was '... too hot for the streams and pastures necessary for human life', while in the northern and southern polar latitudes '... the cold prevents human habitation' (quoted in Martin, 2006: 3). A contemporary version of this narrative is beginning to form around the prospect of rising temperatures in already hot regions of the world. For example, a recent scientific study suggested that wet-bulb temperatures in Arabian Gulf may begin this century to approach a critical threshold of human survivability because of climate change (Pal and Eltahir, 2016). The study generated media headlines such as 'Persian Gulf heat: It may become too hot for humans to survive'[2] and the rather less cautious 'Middle East uninhabitable by 2100'[3], a futuristic re-visiting of Aristotle's torrid zone.

Aristotle was proscribing human habitability not merely through excess heat, but also through excess cold. And here too fears of a changing climate causing glaciers to advance and ice-sheets to grow have recurred throughout the centuries. Stefan Brönnimann (2002) excavated visual imagery from early twentieth-century Swiss posters and magazine and newspaper articles about climate, landscapes and tourism. The motif of advancing Alpine glaciers was often foregrounded in these visual representations of the Swiss environment. As one editor from a 1919 pamphlet remarked, 'Something like an *age-old fear of our ancestors* seems to come to life again: of a global winter that destroyed everything. At the same time one believes that whoever solved the secret of the ice age would be able to understand the magic of today's weather' (Bölsche, 1919: 13, emphasis added). And in the more recent artistic form of film, the climatic extreme most likely to form the storyline for the new genre of Cli-Fi movies is that of the Earth slipping into an ice age and the fears of human extinction that are thus awakened (Svoboda, 2016).

Existential threats to humanity come in many guises. The University of Cambridge's Centre for the Study of Existential Risk – co-founded in 2012 by the former UK Astronomer Royal Sir Martin Rees – adopted the threat of climate-change as one of its five research programmes. This trope of climate change as existential threat has been promoted by other prominent NGOs and public figures beyond James Lovelock and Martin Rees. The parallels between climate-change and the threats to human survival posed by global war have been in circulation for at least 20 years. In 1994, Greenpeace International launched one of its first reports on the impacts of climate change under the title *Climate Time Bomb: Signs of Climate Change from the Greenpeace*

---

[2] CNN (2015) 'Persian Gulf heat: It may become too hot for humans to survive, study warns', 28 October, http://edition.cnn.com/2015/10/27/world/persian-gulf-heat-climate-change/ (accessed 13 May 2016).

[3] A Changing Climate (2015) 'Middle East uninhabitable by 2100', 13 November, https://achanging climate.org/2015/11/13/middle-east-uninhabitable-in-2100/ (accessed 13 May 2016).

**Figure 7.1**   Front cover of Greenpeace International's 1994 report on global warming: 'Climate time bomb: Signs of climate change from the Greenpeace database' (Source: Doyle, 2007).

Reproduced by permission of Greenpeace International

*Database*. The front cover of this report (see **Figure 7.1**) alluded to the threat of a nuclear holocaust by cleverly massaging the image of a yellow glowing sun so that it resembled the mushroom cloud of an atomic bomb.

At the height of the controversy about Saddam Hussein's (eventually fictitious) weapons of mass destruction (WMD) in Iraq, comparisons between the fears evoked by WMD and those of climate-change were made explicit by the leading climate scientist Sir John Houghton. In the summer of 2003, shortly after the invasion of Iraq, Houghton made the comparison explicit, 'The impacts of global warming are such that I have no hesitation in describing it as a "weapon of mass destruction"'[4], while a few years later the physicist Stephen Hawking played to similar fears by claiming that 'Terror only kills hundreds or thousands of people. Global warming could kill millions. We

---

[4] Houghton, J. (2003), 'Global warming is now a weapon of mass destruction', *The Guardian*, 28 July. http://www.theguardian.com/politics/2003/jul/28/environment.greenpolitics (accessed 13 May 2016).

should have a war on global warming rather than the war on terror[5].' Whether through flood or ice, through extreme heat or extreme cold, whether climate change is like an atomic bomb or other weapons of mass destruction, or whether it terrorises populations, human cultures through the ages have found many ways to be fearful of climate and the changes that it wreaks on the human imagination and the physical world.

## Sublime Terror

Whilst climates may be feared for the existential threat they appear to pose if disturbed or destabilised, a different sort of fear is evoked by the idea of the sublime. In western thought the sublime has most commonly been associated with Romanticism and famously inspired by Edmund Burke's 1757 book *A Philosophical Enquiry into the Origin of our Ideas of the Sublime and Beautiful*. Burke saw a contrast between the beautiful, evoking merely feelings of pleasure, and the sublime. The sublime was associated with awe, terror and danger and evoked a near spiritual response:

> Whatever is fitted in any sort to excite the ideas of pain, and danger, that is to say, whatever is in any sort terrible, or is analogous to terror, is a source of the sublime; that is, it is productive of the strongest emotion which the mind is capable of feeling. (Burke, 1968[1757]: 39)

Terror, for Burke, was the ruling principle of the sublime, a sublime which exerted an irresistible pull on the human imagination and especially of its interpretation of the powerful manifestations of nature. The sublime is admired for its overwhelming power and vastness, its ability to shrink the human so it is almost extinguished. In the face of atmospheric vastness and the forces of the wind and weather which impose themselves upon the human soul, it is not mere fear that is experienced, it is a Burkean sublime terror.

This human fascination with awe, terror and danger can be discerned in the encounters with climates at their most extreme. For example, there is a sense of terror evoked by the methane fireballs and the 'hypercanes' of *The Independent*'s front page story with which I started this chapter. There is also a similar alluring terror wrapped up in the exultation displayed by American tornado-chasers and in the vicarious experiences of volatile weather and climatic precariousness made possible through film and the internet. Film critics Phil Hammond and Hugh Ortega Breton allude to it in their review of recent Cli-Fi movies, '... there is often a fascination with [climatic] destruction and annihilation, in which political engagement is implicitly rejected or seen to fail' (2014: 315). This is climate 'red in tooth and claw'.

---

[5] Professor Stephen Hawking, quoted in *The Washington Post*, 'Doomsday clock moves closer to midnight', 17 January 2007, http://www.washingtonpost.com/wp-dyn/content/article/2007/01/17/AR2007011700782_pf.html (accessed 11 June 2016).

It is for similar reasons, I think, that some analysts describe the sensational reporting of weather extremes and the lurid descriptions of prospective climatic disruptions as weather or climate 'porn'. The UK's Institute for Public Policy Research (IPPR) first used the metaphor of pornography in their 2006 report on how climate-change is communicated in language and metaphor (Ereaut and Segnit, 2006). Climate-change in the Anglophone world, they observed, was commonly communicated as 'awesome, terrible, immense and beyond human control, typified by an inflated or extreme lexicon, incorporating an urgent tone and cinematic codes' (2006: 7). Such communicative tropes captivated the attention of the viewer or reader, enticing them through the offer of vicarious encounters with sublime terror. As with the thrills experienced by storm-chasers, the closer one gets to the source of the atmosphere's awe-inspiring power, the more one feels able to transcend the terror that is evoked. It is as though by becoming a part of this powerful event, the observer – in some strange way – becomes the source of it. Whilst captivated by fear, they continue to seek closer union with the storm for the transcendent experience of the sublime that it offers.

## Risk Society

The third context in which I examine climatic fear, beyond those of existential threat and sublime terror, is that offered by Ulrich Beck's idea of *Risikogesellschaft*, or 'risk society' (Beck, 1992). For Beck, increasing prosperity, more intrusive technologies and the increased social complexity and political dependencies of modern states gave rise to this new psycho-social condition. Human concerns in western and other developed nations are now less about war, hunger or political oppression than they are about the pervasive, yet incalculable and often invisible, side-effects of human technologies. Anxiety, fear and dread describe this new human condition, emotions that are most frequently attached to dangers that are yet to be. The reach into the future afforded by powerful scientific predictive modelling brings such putative and unrealised dangers into the present. As Beck himself describes:

> Risk society means that the past is losing its power of determination of the present. It is being replaced by the future – that is to say, something non-existent, fictitious and constructed – as the basis for present-day action... Expected risks are the whip to keep the present in line. The more threatening the shadows that fall on the present *because a terrible future is impending*, the more believed are the headlines provoked by the dramatisation of risk today. (Beck, 1997: 20; emphasis added)

The media dramatisation of risk Beck describes was nicely illustrated in the same year that *Risikogesellschaft* was originally published in Germany. In 1986, the term *Klimakatastrophe* (climate catastrophe) entered the German lexicon. On 11 August, just a few weeks after the nuclear power reactor melt-down at Chernobyl, the front cover of Germany's leading magazine weekly *Der Spiegel* was filled with an image of Cologne cathedral under 10 metres of water. The headline simply read 'Die Klima-Katastrophe', a word subsequently selected by the German Language Society in 2007 as its word of the year.

In relation to climatic fears, risk society manifests itself in various ways. One is through the deployment of the emergency narrative (see **Box 7.2**). By declaring a climate emergency, the prospect is raised that anxieties about a volatile climate can be brought back under human control. But this can only by secured by transferring ever greater powers to the state or, for the community of nations, to the United Nations. 'Emergency' is a risk narrative that changes the terms under which climate might be governed (see **Chapter 10**).

---

## Box 7.2: Climate Emergencies

One of the discursive tropes of Beck's risk society is that of public emergency. The idea of emergency forges a distinction between the normal state of affairs, which is broadly manageable using the political institutions of the modern state, and the unpredictable irruption of emergencies which threaten to overrun these same institutions. The ubiquity of emergencies in today's world therefore becomes a strategy for retaining public faith in the capacity of modernity to manage the world. If the normal organs and reach of the state are deemed insufficient to deal with danger, then declaring an emergency legitimises additional state powers to re-establish normality. In today's culture of risk and emergency it is not surprising that ancient anxieties about disorderly climates now surface as calls to consider, or even to declare, a climate emergency. Consider these recent examples: 'Climate is nearing dangerous tipping points. Elements of a "perfect storm", a global cataclysm, are assembled'[6]; 'Our world is far beyond dangerous interference with the climate system, in a state of worldwide emergency'[7]; '... climate emergencies, such as melting Arctic sea ice and polar ice sheets and a food crisis in the tropics [may not be preventable]' (McCusker et al., 2012: 3096).

But who is to convert diffuse and intangible fears about some future (or present) climatic condition into the formal category of a climate emergency? Emergencies can only be declared; they cannot be discovered. The declaration of an emergency is therefore a political act, an exercise in power, and will inevitably be used for political purposes. By definition,

*(Continued)*

---

[6] James Hansen, Testimony to the US Congress, 23 June 2008, 'Global warming twenty years later. Tipping points near', http://www.columbia.edu/~jeh1/2008/TwentyYearsLater_20080623.pdf (accessed 11 June 2016).
[7] Climate Emergency Institute website, http://www.climateemergencyinstitute.com/cei_mission.html (accessed 13 May 2016).

*(Continued)*

emergency situations are extraordinary and exceptional. Declaring an emergency invokes a state of exception which carries many inherent risks for a democracy: the suspension of normal governance, the use of coercive rhetoric, calls for 'desperate measures,' censored deliberation, militarisation (Sillmann et al., 2015). In the case of climatic emergency it is not at all clear how such an emergency would be defused. Some commentators argue that in response to future climate emergencies it may be necessary to develop novel technologies for engineering the world's climate through deliberate intervention (see **Chapter 10**). Others see the very declaration of a climate emergency as tantamount to an anti-democratic power grab. As sociologist Craig Calhoun remarked a few years ago, 'The production of emergencies, and the need to address them, has become one of the rationales for assertion of global power' (Calhoun, 2008: 379).

Another manifestation of risk society is the normalisation of fear as an appropriate emotional response to the climatic future, rather than as a reaction to existing or past climatic disasters, as in previous eras and cultures. Under the conditions of risk society, the mediatisation of prospective climatic changes induces heightened feelings of a vague dread and anxiety about the future. What, for example, is an appropriate emotional response to reading the front page of *The Independent* newspaper shown earlier? Beyond scepticism or a sense of the sublime, the most likely emotion to be released is that of fear. Will I or my children be among those 'few survivors eking out a living in polar refuges'? In a study of Swedish citizens' reactions to media headlines of prospective future climates, Birgitta Höijer (2010) revealed the range of emotional responses to these verbal and visual cues. Fear was the most predominant. She found the abstract future risk calculus offered by science was objectivised by the media through representing climate risks as visual phenomena existing in the physical world, for example, through specific storms, heat waves or floods. Popular media represented climate-change as an impending threat that is seriously, not trivially, bad, but also as one that people sense they have no power to affect. As Beck surmised in his risk society thesis, negative and fearful emotions are thereby released.

A further characteristic of risk society, argued by some as being provoked by such emotional reactions, is a deterioration in mental health. UK psychotherapist Rosemary Randall has formally identified a psychological condition she calls 'climate anxiety'. Portrayals of climate-change can appear overwhelming, producing 'monstrous and terrifying images of the future accompanied by bland and ineffective proposals for change' (Randall, 2009: 118). Rather than the experience of climate terror as sublime, for some people climatic disordering triggers feelings of despair, helplessness and anxiety. Randall argues that psychoanalytic models of grief and loss can help understand and treat these symptoms of the climatically anxious. Facing and mourning the imagined losses associated with climate-change is the first stage of this treatment. Such emotions need inserting into public narratives of a changing climate, she claims, a pre-condition for releasing the emotional energy needed for psychological rehabilitation.

## Chapter Summary

Ecological disorder and fear of the unknown future are enduring sources of human anxiety. In this book I have developed an argument that offers climate as an *idea* invented by people to help stabilise one form of physical chaos – the weather – in order for them to live stably and creatively amidst this uncertainty. In this chapter I have explored what happens to human emotions when this stabilising idea becomes destabilised; when either the experience of past climatic disorder or the claims of a future descent into climatic chaos feeds the imagination in powerful ways. And as I have shown in **Chapter 3**, fears prompted by climatic disturbance can be either released or contained depending on which sources of climatic knowledge are deemed trustworthy by different individuals and cultures.

Yet although fearful interpretations of climatic behaviour are common throughout human history, these fears are always mediated culturally, through myths and tropes, through language and metaphor or through powerful media or political interests. For example, the contrasting metaphors of 'climatic stability' and climate 'tipping points' conjure very different imaginative worlds. Both are in circulation, yet people respond to these manifestations of putative order/disorder in different ways. Those inclined to see natural systems as innately fragile and easily perturbed are more likely to see looming existential threats as climates are perceived to change. For others, perhaps with different cultural, spiritual and imaginative dispositions, climatic dangers may be diverted into experiences of the sublime or attenuated through faith in a powerful deity. Climatic fears are bound up in wider cultural narratives of apocalypse, risk society, emergencies and psychological (in)security. Climate chaos exists in the imagination as much as it can be discovered through scientific instrumentation and calculation. And it is to the cultural practices of climate representation that I turn in the next chapter.

## Further Reading

Bruckner, P. (2013) *The Fanaticism of the Apocalypse: Save the Earth, Punish Human Beings.* Cambridge: Polity.

Boia, L. (2005) *The Weather in the Imagination.* London: Reaktion Books.

Skrimshire, S. (ed.) (2010) *Future Ethics: Climate Change and Apocalyptic Imagination.* London: Continuum.

Spratt, D. and Sutton, P. (2008) *Climate Code Red: The Case for Emergency Action.* Melbourne: Scribe Publications.

# 8

# Representing Climate

## Introduction

I described the paradox of climate in **Chapter 1**, namely that while you and I both live *in* climates I cannot *show* you this climate in which we live. I can feel the force of the wind or the grip of the cold on my body – these are the direct, unmediated sensations of weather acting upon me. But I cannot 'see' or directly experience my climate, neither the climate of the place in which I live nor, even less, the climate of the planet. This then constitutes the representational problem of climate. Climate is an intuitive idea, familiar to all human cultures, which helps make sense of the world. And yet, while intuitive, it is an idea which can only be made tangible – made visible – through the creative and political work of representation.

Philosophers, scientists and artists through the ages, and around the world, have all grappled with this problem of representation. Claudius Ptolemy's map from second-century (CE) Alexandria, which depicted seven latitudinal *klimata*, was one of the earliest formal representations of climate. In 1817 Alexander von Humboldt invented the isotherms that joined together in cartographic form disparate places which shared a common climate (in his case average temperature), yielding what some have called the first climate map. And in the early twentieth century, the Russian geographer Vladimir Köppen developed his classic world map of climatic zones based on the view that different climates are best revealed through characteristic surface vegetation (this idea was the inspiration for the much later Eden Project in southwest England, a physical construction which encapsulates Köppen's bioclimatic zones in a series of plastic domes). Artists too have participated in the ambiguities of climatic representation, for example, the Romantic artist John Constable and his depiction of English skies and the French Impressionist Claude Monet and his evocation of London's late Victorian climate.

A more recent venture into this territory has been the creation in northern Germany of a museum and public education centre in Bremerhaven called Klimahaus, or Climate-House. Partly taking inspiration from the Eden Project's bioclimatic domes, Klimahaus invites visitors to 'Join us for a fantastic journey through our climate'[1]. Visitors are taken through nine 'rooms' covering various climatic regions of the Earth, linked through their common geospatial attachment to Bremerhaven's degree of longitude, 8°E. At each stop in the journey, visitors enter an exhibition designed to visualise the climate of a different latitude on this meridian. These exhibits and artefacts are suggestive of climate's role in shaping distinctive biological life and human cultures. Evocative physical elements of each respective climate – heat or cold, humidity or aridity; these are stable conditions since there is no varying weather in Klimahaus – are regulated through air-conditioning. Visitors are introduced to manicured human customs and traditions which, too, seem stable and timeless caricatures of weathered cultures.

Klimahaus[2] is a bold and interesting attempt to 'perform' the idea of climate, taking inspiration from Ptolemy's zones, von Humboldt's isotherms and Köppen's bioclimatic ecologies. Visitors are taken out of their own climates and immersed in a series of technologically controlled interior atmospheres. Klimahaus tells a story of how climate both divides and connects humans. On the one hand, climate is shown to explain difference – in both ecologies and cultural practices. But the underlying theme of Klimahaus is also one of connection. Although the world's regional climates are first isolated and manufactured, they are then gathered together, warehouse-like, in 'a gracefully bulbous glass and steel construction'. The Klimahaus therefore represents the diversity of the world's climates in a single space, an experience to be enjoyed during a single afternoon. Furthermore, the museum issues a political injunction that each visitor has a common duty to protect these climates from further deterioration. The message of the museum is that 'climate protection is everybody's business!' Visitors are invited to open their own 'climate account' and discover in an easy way the 'day-to-day opportunities there are to live more climate-consciously'. As Mahony concludes his essay, 'Klimahaus may be considered a paragon of twentyfirst century forms of globalisation'.

\* \* \*

As I argued in earlier chapters, climate dwells in the human imagination as much as it is discovered in the material world. Artistic representations of climate give expression to the former, while scientific investigations pursue the latter, but either way there is no unmediated access to the human experience of climate and

---

[1] Klimahaus, http://www.klimahaus-bremerhaven.de/en/home/home.html (accessed 13 May 2016).
[2] This account draws partly on a short essay about Klimahaus written by Martin Mahony in July 2015, see http://thetopograph.blogspot.co.uk/2015/07/klimahaus-bremerhaven-in-world-interior.html (accessed 13 May 2016).

its changes. The forms of cultural expression illustrated in this chapter are not simply adjuncts to climate science, trying to communicate the essential truth of climate which science has dutifully unveiled. Nor are these representations (always) designed to coerce audiences to alter their behaviour in certain pre-defined ways to atone for, or to alleviate, disruptions in climate. Rather than nudging changes in lifestyles, the purpose of different cultural representations of climate is more often to provoke deeper reflection and self-understanding about the world in which humans find themselves. Analysis of such representations shows how the idea of climate works to stabilise cultural relationships between humans and their weather.

The chapter explores ways in which the idea of climate and its changes has been represented, first, in scientific practices and then in visual, literary and performance cultures. The various techniques of representation illustrated here include modelling, animation, visual art, photography, creative fiction, theatre and film, with examples drawn from around the world.

## Scientific Mediation

Science has been an extraordinarily successful human enterprise for under-standing the physical processes which form, connect and transform the material world. Through ingenious instrumental devices, scientists are able to see both the microscopically small and the unimaginably vast. Yet privileging the visual in science communication has proven problematic. If 'seeing is believing', then giving visual form to the idea of climate (and its changes) becomes a pre-condition of belief in climate (and its changes). But there is no simple instrument, nor a complex one, that extends the human senses such that climate can be witnessed first-hand. I mentioned earlier the techniques used by Ptolemy, von Humboldt and Köppen to bring to life in cartographic form the idea of climate. The climate sciences of the last 50 years have grap-pled with a greater problem: how to visualise for diverse audiences the vast and still growing Petabytes of digital weather, oceanographic and virtual (simulated) data, which together construct the notion of climate in the scien-tific imagination (see **Box 8.1**).

---

### Box 8.1: Virtual Climates

New digital visual technologies combined with the data-storing capabilities of 'the cloud' are transforming the way climates can be visualised. Computer animations of the planet using Google Earth, and their accessibility to ever greater swathes of humanity, are changing what sort of object the world is imagined to be. They are changing the possibilities of how climates, past and future, are made visual. Through these animations climate 'becomes both manufactured and remediated media' (Gurevitch, 2014: 90). Climate is not just 'out there' or in the mind. Climate is now also in the machine. Future climates can be visualised and re-engineered so easily through these animating technologies that the feasibility, even the desirability, of a future climate-engineered planet becomes normalised (see **Chapter 10**). The desire for climate control, previously achieved in homes, buildings and cars, can now find consummation in a thermostat-manipulated planet, brought to life as it were through Google Earth's digital platforms. How climate is represented changes the imaginative possibilities of how people might live with, or alter, their climate.

Other mega-corporations beyond Google are also taking ownership of the means of climate representation. In recent years, the corporate giant Amazon has partnered with NASA to offer an internet-based digital platform (marketed through Amazon Web Services) which allows users to visualise climates, from global to local scales. Such a platform 'seamlessly integrates' large empirical climate datasets, satellite imagery of the world's weather and computer model simulations of past and future climates (all supplied by NASA). It too offers new possibilities for manipulating climate and making it visible. As an Amazon 'Public Data Set', these digital data allow citizens and governmental, business and humanitarian interests to map, plot, colour, redesign and interrogate climates in real time. This is climate 'democratised' or 'corporatised', perhaps both.

There are no self-evidently correct or unambiguous ways in which these climatic data *should* be represented in visual form. As with artists, there may be some conventions to follow, but there is no rule book; creativity and innovation rather than imitation are the mark of the true scientist (and artist). All scientific images of climate are artificially constructed, even if – through repeated public circulation – they become reified as seemingly literal representations of a physical reality. Decisions by the scientific (visual) designer are required from start to finish: which cartographic projection to use; what numerical interval and colour scale to adopt; which variables to highlight, which to downgrade? Some of these choices may be shaped through cultural conventions, for example, the red of a rising temperature curve or the deep crimson colouring of computer generated maps of future climate. As visual analyst Birgit Schneider points out, 'An expert visualiser may simply see the colour red as the only intuitive way to code heat symbolically, yet it might at the same time provoke imaginations of the apocalypse of an inevitable climate catastrophe' (Schneider, 2012: 186). But choices remain and they convey different epistemic, emotional and political messages, with significant repercussions for how the resulting visualisations of climate are received and interpreted.

Scientific representations of climate become political in other ways also. Given the vast volumes of climatic data generated by observation and simulation technologies, visual techniques have to select judiciously from among the vast array of data options. As much as scientists design climates to be 'visible', they are also deciding which aspects of climate are to remain 'invisible'. Exclusions are as important as inclusions. It was in such exclusionary practices that a number of climate scientists were accused of participating in the 'Climategate' affair of 2009. Some of the released email correspondence between leading scientists revealed the apparent intention of dendro-climatologists to 'hide the decline' when designing a high-profile graph to show long-term changes in hemispheric temperatures. While such selective practices were deemed nefarious acts by critics, in fact 'hiding' data is an essential skill deployed by any good visual science communicator. The visible climates 'seen' in public are the end result of complex processes of measuring, manipulating, simulating and designing. But the arguments and negotiations between experts take place back-stage, away from the public gaze, even though – as with creative works of art – it is sometimes the arcane and untidy processes of *producing* a work of representation that are as interesting as the final presentable *product*.

Visual representations of scientific data – of climate as much as of other objects and concepts – do not offer transparent windows on reality. Ignorance of this fact is worrying, as pointed out by rhetorician Lynda Walsh (Walsh, 2015). If climate scientists view the selections and rhetorical skills they deploy as distortions of a pure reality, these then become practices which are to be hidden or denied. On the other hand, Walsh's concern for the non-expert is different. Either scientific representations of climate are too difficult to decode by the untrained mind, in which case public understanding ceases, or else, she points out, non-experts *can* decode the graphic, at which point they sense they have appropriated a literal depiction of climate. When others subsequently point out the value-based decisions and judgements which scientists used to construct their graphic or map, such people interpret this as an attack on the integrity of science. A firmer truth to grasp, for expert and non-expert alike, is that there can be no unmediated access to climate.

It is not only through data visualisation that science represents publicly the idea of climate. Another way in which scientific understandings of climate are brought into cultural circulation is through naming conventions and, in particular, through personification. Perhaps the best known example of this latter technique is the climatic phenomenon of El Niño (and its associate La Niña). It was Peruvian fishers in the late nineteenth century who first gave name to the warm ocean current appearing occasionally off the coast of Peru. This ocean warming brought welcome rain to the arid coastal desert and since it typically emerged in the months leading up to Christmas, they christened the phenomenon 'El Niño' (in English 'the little boy'), metaphorically the Christ-child. Scientists first appropriated this nomenclature in the 1920s and today it acts as a powerful representation of a complex set of regional climatic characteristics. Around the world today, El Niño represents both scientific and cultural understandings of diverse climatic phenomena.

Another example of climatic personification is of violent storms. Since 1950 the US National Hurricane Center has named hurricanes alphabetically using alternating male and female names. Whilst such personification of natural climatic events carries certain benefits for public communication, recent research also shows how such representations of climate cannot escape mediation through culture. Since 1950, female-named hurricanes have produced almost double the number of fatalities in the United States than have male-named hurricanes, on average 45 deaths compared to 23 (Jung et al., 2014). The explanation for such an outcome is that people's behaviour in the face of hurricane warnings is influenced by gender-based expectations. American citizens associate femininity with less danger and risk than they do masculinity. When warned of a 'female' hurricane they fail to take the same precautions as they would with a 'male' hurricane. Not only can there be no unmediated access to climate, but all forms of representation are interpreted through culture.

## Visual Arts

I have briefly summarised some of the techniques and tensions which scientists use and encounter as they seek to represent climate in public communication. Other forms of cultural representation beyond science offer different opportunities for enlivening public understanding and engagement with the idea of climate, while also encountering representational difficulties of their own. Artistic practices offer creative ways of 'making climate visible' which can accommodate the ambiguities and contradictions of knowledge and experience which science seeks to eradicate. The idea of climate can therefore be approached through multiple rationalities and viewed in different cultural registers than simply those offered by science. And when it comes to representing the idea of climate-change, art is a less didactic form of communication than is science. The artist encourages the viewer to reflect on their own understanding of climate and its changes and, through the imagination, what this might mean for them.

One way in which climate has been studied through paintings has used the idea of landscape meteorology. The meteorologist L.C.W. Bonacina gave form to this idea in his 1939 essay, 'Landscape meteorology and its reflection in art and literature'. The artistic side of meteorology, he explains, is 'those scenic influences of sky, atmosphere, weather and climate which form part of our natural human environment ... whether ... carefully stored in the memory, or as photographed, painted or described' (Bonacina, 1939: 485). The problem here is that visual artists, unlike climate modellers, do not usually start out with the intention of bringing physical climates into clear optical focus. It is hardly likely, for example, that early modern Dutch landscape artists such as Bruegel or Avercamp were seeking to faithfully represent the slow-cycle changes in climatic conditions experienced in northern Europe during the 'Little Ice Age' of the sixteenth and seventeenth centuries. Using historical paintings as a form of 'climate archive', as a natural scientist would use tree-rings or lake varves as a proxy for past physical climate, is untenable – even if it has occasionally been attempted (e.g. Neuberger, 1970). The historical contingencies of painting – technical issues such as the availability of colours and

changing social and religious cultures – mean that artwork has an insecure
foundation as a proxy for physical climate.

What is intervening between the physical manifestations of climate and their
cultural representation is the human imagination and its use of perspective. The
concept of 'Realism' was an invention of nineteenth-century France – at least as
a self-conscious art movement; recent art theorists such as Erwin Panofsky suggest realist perspectives can be traced earlier than this. Realism sought to
'represent things as they are ... supposing that I (the perceiving subject) did not
exist' (Rubin, 1996: 53), but it paved the way for Impressionism, which offered
a different way of seeing the world to that of classical forms of representation.
Impressionists sought to capture the transience of light and shade through the
use of new brush strokes and colour palettes. Impressionist Claude Monet has
been regarded by some as a 'climate artist' (Thornes and Metherell, 2003).
Monet's depictions from the 1890s of London's late Victorian climate remain
some of the most evocative representations left to us of the peculiar twilight
climate of London fog, smoke and atmosphere (see **Figure 8.1**). These are representations of what amateur meteorologist Luke Howard 60 years earlier had
neologized as London's 'urban climate'. Monet's visual representations of particular atmospheric scenes contributed in no small part to the symbolism of
power, mystery and prosperity which London's fog-saturated climate extended
over cultural imaginations of the first half of the twentieth century.

Photography is a contrasting form of representation to that of canvas and
brush, but one that is also circumscribed by the intangible nature of climate

**Figure 8.1**  'The Thames below Westminster'; by Claude Monet, 1871. Monet has
been described as the first 'climate artist'. Oil on canvas, Collection Lord Astor of
Hever; © National Gallery, London.

and by the cultural and material influences on its practice. If photography is understood as offering a mimetic technique for faithfully capturing the material traces of objects, landscape or people, then photographing climate implies that climate must be 'there' to capture on film. But photography cannot be understood this way (Sontag, 1977) and neither can climate. This means that approaching photographic representations of climate and its changes as conveyors of reliable truth is problematic. The frustrations and paradoxes of climatic photography as a form of representation were encountered by Greenpeace International in its 1990s public campaigns to make climate-change visible and therefore 'real'. Julie Doyle's investigation of these efforts concludes that the photograph '... as a powerful tool of environmental persuasion and documentation, is inscribed by the representational limitations of the visual as a discourse of seeing and truth' (Doyle, 2007: 147).

## Literary Fiction

The idea of climate has been represented in multiple and complex ways in literary fiction. At one level, weather (more so than climate) has functioned ubiquitously for writers as a way of introducing emotion and sentimentality into their narratives, what John Ruskin in the nineteenth century referred to as 'the pathetic fallacy', or emotional falsity. Thus the storm outside the window complements the heroine's tempestuous emotions, or intense summer heat discloses erotic passion as, for example, in L.P. Hartley's 1953 novel *The Go-Between*.

In terms of literary genre, ancient religious and apocalyptic narratives developed fictional accounts of wayward and dramatic climatic behaviour and its effects on human culture, such as the Epic of Gilgamesh or the Apocalypse of John (as noted in **Chapters 4** and **7**). Science fiction is another genre which actively engages with climatic representation, in this case climates of the future or of other planets. Often these depictions are of dystopic climates or else of deliberately re-engineered habitable climates. These latter reflect the idea of 'terraforming' (a precursor to climate engineering), a term introduced by the writer Jack Williamson in his series of short stories in the 1940s (Trexler and Johns-Putra, 2011). For example, Kim Stanley Robinson's 'Mars' trilogy from the 1990s details the colonisation and gradual transformation of the Martian climate over hundreds of years to render it fit for human habitation.

Science-fictional ('Sci-Fi') climates have more often been dystopic than utopic, especially when earthly climates are in view and especially in writings from the last 50 years as public concerns about environmental change have grown. But the causes of such dystopian climates have shifted from being natural – as in J.G. Ballard's *The Drowned World* (1962) – to human-caused, as in Arthur Herzog's *Heat* (1977). And it is the idea of climate-change which in the last decade has led to a burgeoning of fictional work dealing with the idea of climate. Rather than climate acting as a mere 'setting' for story, these fictions seek to explore the connected wider social, political, cultural, economic and technological processes which lead to climate's unsettling. An early exemplar of this trope was Robinson's 'Science in the Capital' trilogy – *Forty Signs of Rain* (2004), *Fifty Degrees Below* (2005) and *Sixty Days and Counting* (2007) – in which science and political will unites in order to redress disaster brought about by human-caused climatic change. Such literature develops a new version of the pathetic fallacy, writers now using the idea of climate-change as a synecdoche for a dysfunctional society. These new works of climate fiction have in recent years been labelled 'Cli-Fi', although whether Cli-Fi is a distinctive genre is debatable. Literary genre is always fluid in nature and many works of fiction which deal with climate straddle generic boundaries (Johns-Putra, 2016). Many different genres work with the idea of climate-change, whether science fiction, dystopia, fantasy, thriller or romance. Cli-Fi is not necessarily a self-standing genre.

Representing the idea of climate and its changes in fiction is undoubtedly challenging, but it is a challenge being taken up by a growing number of writers. What so-called Cli-Fi novels have in common is that they grapple with the complexity of climate-change as a co-constituted cultural *and* scientific phenomenon. Representing climate in fiction is never an exercise in the mere translation of scientific concepts into literary form. It is rather about recognising the physical, sociological, political and psychological complexities of climate and capturing these complexities in a story. Some ecocritics read this new climate fiction simply as a way of understanding these cultural complexities. Others suggest that Cli-Fi provides readers with an illustration of how to live with a changing climate. For others still, it has spawned a new form of ecocriticism called 'critical climate change' (Johns-Putra, 2016), exemplified by reactions to English novelist Ian McEwan's well-trailed satirical allegory *Solar* (2010) and German author Ilija Trojanow's novel *Melting Ice* (2011). *Solar* engages with the cultural politics of climate-change, but many critics found the novel and its characterisation disappointing. According to Garrard (2013), *Solar* reveals the extent of the challenge of handling climate-change in 'realist' novels, as opposed to its place in futurist or science fiction novels where more degrees of freedom can be used to engage the imagination. Critics of *Melting Ice* also draw attention to the aesthetic challenges of writing about climate-change, but this novel also reveals the tension between the novelists' confessional and didactic impulses when writing about climate-change.

Whatever the critics' reaction, literary representations of climate seek to accomplish what all good writers of fiction aim for: to draw readers into new imaginative worlds in which they learn more about themselves and their emotional, intellectual, philosophical and spiritual capacities. Representing climate through fiction is about opening up new possibilities for readers to feel, taste, smell and imagine climate and its multiple meanings. Cli-Fi is a way of humanising climate. Whatever else it accomplishes, 'the diverse storylines of climate fiction make it impossible to think about the future in a singular way' (Milkoreit, 2016: 179).

# Performance and Film

Beyond visual art and literary fiction, the visual and the textual can be brought together to represent the idea of climate through performance: in theatre, dance, song and film. Staged dramas are usually performed in the controlled, artificial environment of the theatre and, if successful, offer gripping insight into comic or tragic human behaviour. These contrived and ancient forms of story-telling may seem an unusual way of representing climate, but if, as I am arguing here, human culture is at the heart of the idea of climate then theatrical performance might offer powerful insights into the social drama of climate (Smith and Howe, 2015). The ability of cultural performances to challenge stereotypical representations of the human consequences of a changing climate is well illustrated by *Moana: The Rising of the Sea*, a music-dance-drama originating from the Pacific islands (see **Box 8.2**).

---

## Box 8.2: *Moana: The Rising of the Sea*

One of the inevitable consequences of a warming global climate is a rising of the oceans. But as with climate, rising seas are imperceptible to human senses and can only be represented through mediators. Part of the familiar visual vocabulary of changing climates and rising sea-level is the Pacific island atoll and the stranded helpless island victim forced to migrate and in need of 'saving' by an enlightened world. For western and northern media especially, such island images function as attractive and ubiquitous representations of climate-change. They convey a sense of (heightening) drama, anchored in the present and made visible through images of an 'island paradise' under threat. Yet such representations are deeply rooted in a particular western cultural imaginary and perpetuate colonial and Eurocentric constructions of Pacific islands. These western representations of Pacific climates and cultures are based on three principles: insularity, alterity and concretion (Kempf, 2015). They conform to a particular western conception of climate-change: that the consequences are far away (insularity); that it is 'other' people who are affected (alterity); and the need for some graspable, physical reality through which climate-change is decisively revealed (concretion).

Yet there are other ways of representing climate and rising seas in these places beyond those originating in western imaginations. One expression of such representational resistance from the Pacific is the elaborate Oceanian music-dance-drama, *Moana: The Rising of the Sea*, written by Fijian playwright Vilsoni Hereniko with a music score by Igelese Ete[3]. Commissioned in 2013 by a group of anthropologists, *Moana* was first performed publicly

*(Continued)*

---

[3] A short film version can be found here: https://vimeo.com/111959080 (accessed 11 June 2016).

*(Continued)*

in Fiji in December 2013 and 18 months later embarked on a mini-European tour, encompassing Norway, Scotland and Belgium. By offering an account of rising seas which is rooted in Oceanic cultures, assimilating narrative, myth, dance, music and video, *Moana* challenges western notions of help-less victimhood. It is designed to shake audiences out of their comfortable certainties regarding the place of climate in contemporary Pacific politics. *Moana* does not merely draw attention to the challenges faced and under-stood by Pacific islanders themselves. It inverts the power gradient about the representation of climate which currently extends *from* the west *to* the Pacific. It rests on the rich cultural resources of 'our sea of islands', Epeli Hau'ofa's powerful concept of identity which binds together the peoples of the Pacific (Hau'ofa, 2008). *Moana* returns agency to the peoples of the Pacific. It says to the world: '*This* is what climate-change means to us. We will represent ourselves on the global stage; do not do this for us'.

The virtue of stage over film is that it offers a setting for debate and dialogue in contrast to 'the restless visual vocabulary of film' (Bottoms, 2012: 340). The stage offers a suitable vehicle for representing the complexities of the emotional entanglements with changes in climate and the human fears and aspirations these entail. Dramatists may still wrestle with the challenges of finding appropri-ate theatrical forms through which to address these issues, but a growing number of climate plays have been written and performed. Four of these appeared on the London stage during the period 2009–2011: *The Contingency Plan*, *The Heretic*, *Greenland* and *Earthquakes in London*. Building on a long British tradition of politically engaged drama, these plays dramatised the often difficult relationships between scientists, politicians, activists and lay publics. How climate-change was mediated to the public through these stage perfor-mances revealed the playwrights', and therefore the audiences', conflicted relationship with the idea. Each play grappled with the cultural politics of climate and the unruly boundaries between scientific truth, personal integrity and cultural belief. Ironically the play that sought hardest to police the truth-fulness of climate science – *Greenland* – was the play that for the critics was least successful dramatically (Bottoms, 2012).

Representing climate through film offers different possibilities than does the stage. Over 60 films of fiction have taken the idea of a changing climate as their central motif and these have been analysed in terms of genre, plot and characterisation (Svoboda, 2016). Most of these post-date the Hollywood blockbuster movie *The Day After Tomorrow* (TDAT) released in 2004, but none have been as successful in terms of their box-office appeal. By exploiting the ancient trope of the mythical great flood, TDAT developed a variation of apocalyptic discourse connected to a prospective radical change in global cli-mate. However, rather than motivating individual or collective responses in the face of impending disaster, a goal of classic apocalyptic narratives (see **Box 7.1**), the rhetorical structure of the film dissipates such response. 'TDAT provides a world in which all the answers are known ... and there is little left to do but wait for and survive the purification brought by nature's retribution' (Salvador and Norton, 2011: 60). In contrast, Jeff Nichols's 2011 film *Take Shelter* deals with new forms of unsettling and uncertainty that the idea of climate-change elicits. The film challenges the conventional epistemology by which a changing climate is made known – that of science – by juxtaposing different rationalities, such as dreams and filmic transgressions of sound and image. These representations open up more imaginative spaces for audiences to reflect on the unsettling of certainties which the idea of climate-change provokes. *Take Shelter* is a good example of the creative and subversive possibilities of representation using film media.

A very different filmic representation of climate was Al Gore's 2006 documentary movie *An Inconvenient Truth* (AIT). This has been one of the most closely scrutinised climate films, partly because of the salience of its presenter, but it never achieved the popular appeal of TDAT. (Nevertheless, an operatic version of AIT was performed on stage in La Scala in Milan in 2011, although as one opera critic remarked beforehand, 'Al Gore's political life may not have enough core dramatic elements – sex, betrayal, murder – to sustain the plot'!; Anon, 2008.) In AIT climate is represented in sublime, Romantic and pastoral incarnations, but also in terms of wild and threatening danger. Communication specialists Rosteck and Frentz suggest that the 'bifurcated environmental rhetoric' (2009: 16) of AIT captured the diverse cultural engagements with the idea of climate, which may be as contradictory as Gore in the movie reveals his own to be. Different cultural representations of climate usually tell us more about ourselves than they educate us about a 'world out there'.

## Chapter Summary

Climate is an imaginatively fruitful idea and so it is inevitable that it will be represented in different ways. There is no single timeless truth about climate and what it means for people waiting to be revealed through science or through art. Different meanings of climate are constructed and communicated through representations and it is through these that human stories and artefacts become 'weathered'. Cultures become shaped by the idea of climate and, in turn, transform what climate is understood to mean. The multiple meanings of climate are therefore constructed in public, culturally, drawing upon proliferating and at times competing scientific and artistic representations.

Nevertheless, I have shown that representing the idea of climate is a challenge both for scientific practice and for the arts. Climate and its changes are beyond mere mimetic representation, whether by, for example, computer simulation or photography. And since there can be no unmediated access to climate, all representations of climate are in the end political acts. That is, they are engaged in constructing different and selective climatic realities: material, ideological, imaginative, normative. Inclusions and exclusions are inescapable and ideologies become embedded in all forms of representation, whether scientific inscriptions, visual vocabularies, stage performances or literary texts. Emanating from these forms of representation are invitations to see the climate this way or that, to lay blame for (or to take responsibility to solve) climatic ills in particular ways.

A question frequently asked about contemporary climate art is therefore whether it has the power to change beliefs, attitudes and behaviours in ways which scientific representations of climate-change seem not to possess (Miles, 2010). But the question is badly posed. Art that is inspired by the idea of climate is just as likely to problematise presumed pre-existing solutions or attitudinal changes as it is to persuasively advance specific solutions to a pre-existing problem. Despite frequent exhortations to the contrary, climate science cannot 'demand' of people any particular course of action, and neither can art. It is not didactic. It cannot instruct people in sustainable, just or alternative living. But good climate art will engage human faculties to provoke reflection on the profound questions prompted by a change in climate: the good life to be admired, the future to be aspired to and the responsibilities they have to others, both human and non-human.

*  *  *

In Part 3, 'The Futures of Climate', I consider the ways in which the futures of climate and humanity are inescapably bound together. This occurs through human enterprises of climate prediction (**Chapter 9**) and the putative desire to redesign climate on a global scale (**Chapter 10**). In each case, questions are raised about how people seek to govern their climates and what in fact is being governed in the name of climate (**Chapter 11**). Finally, I end the book with some reflections on possible futures of *the idea* of climate and what it may or may not yet accomplish (**Chapter 12**).

## Further Reading

Doyle, J. (2011) *Mediating Climate Change*. Farnham: Ashgate.
Schneider, B. and Nocke, T. (eds) (2014) *Image Politics of Climate Change: Visualisations, Imaginations, Documentations*. Bielefeld, Germany: Transcript Verlag.
Smith, P. and Howe, N. (2015) *Climate Change as Social Drama: Global Warming in the Public Sphere*. New York: Cambridge University Press.
Trexler, A. (2015) *Anthropocene Fictions: The Novel in a Time of Climate Change*. Charlottesville, VA: University of Virginia Press.

# Part 3
## The Futures of Climate

# 9
# Predicting Climate

## Introduction

The art of weather forecasting has many convoluted and colourful histories, many of them inspired by the desire to foretell general human fate as much as the future of the skies. There is also a long cultural history of claims-making about the *climatic* future. Prophecies and predictions of future climates have been issued in all cultures, based on different understandings of the causes of climatic change (see **Chapter 4**). I have already drawn attention to some of these claims in earlier chapters, for example the myth of Noah's Flood, the Hollywood movie *The Day After Tomorrow* and narratives of an impending climatic apocalypse. Strictly scientific predictions of future climate, however, are more recent in origin, yet they should be seen as being only the latest in a long tradition of prophetic forms of climate knowledge. Scientific predictions of climate have come to rely almost exclusively upon models of the climate or the Earth system (see **Chapter 3**). These models are formal mathematical representations of the fundamental physical processes which govern the flows of energy, mass and moisture between the physical components of an interconnected planetary system.

With climate understood *this* way, as an interacting physical system, models can be used to simulate the climate of any planet, whether real or imagined, so long as the basic configurations of land and ocean, and their height and depths, are known. Such reasoning has been used by one group of scientists to predict the climates of Middle Earth, the land invented by novelist J.R.R. Tolkien and in which the tales of *The Hobbit* and *The Lord of the Rings* are set. During his professional life, Tolkien developed a sophisticated geography, history and culture of this imaginary world. This was a world in which weather played important roles in his narratives, whether the dark and violent storms of Mordor, the capricious fog which fell on Frodo and Sam on the Barrow Downs or the warm balmy breezes of the forest realm of Lothlórian. But what of the *climates* of Middle Earth, the systematic differences in patterns of weather across this land?

Scientists from the University of Bristol in England, led by the wizard 'Radagast the Brown', used a variant of the UK Met Office Hadley Centre's climate model to predict the climates of Middle Earth during the Second Age of Arda (Lunt [Radagast the Brown], 2013)[1]. Based on Tolkien's geography (according to *The Atlas of Middle Earth*; Fonstad, 1991), the climate of The Shire – the home of hobbits Bilbo and Frodo Baggins – was simulated to be very similar to the physical climate of the counties of Lincolnshire and Leicestershire in the English East Midlands. It was also similar to the climate of the region around Dunedin in the South Island of New Zealand, not far from where many of the scenes of *Lord of the Rings* were filmed in the screen adaptations. Both regions have temperate climates and rich vegetation, but without great extremes of heat and cold, drought or rain. In contrast, Mordor – the land of the evil Sauron – emerged with a hot and dry climate, similar to that of Los Angeles, western Texas or Alice Springs in central Australia. These inhospitable climates yield little vegetation and are more exacting for hobbit and human civilisations alike. Radagast the Brown's climate predictions also explained the existence of a dry climate east of the Misty Mountains and why it was from the Grey Havens, with the prevailing winds blowing from the east, that the elves set sail for their journey to the west.

These predictions of Middle Earth's climate point to Tolkien's perceptive intuition of the dynamics of how physical climates come into being. They also confirm his mastery of the art of 'sub-creation' (Flieger and Anderson, 2014); worlds brought into being through the human imagination which weave together geography, climate, language and myth to construct powerful and engaging narratives which speak of a deeper reality. In contrast to the power of Tolkien's imagination, Radagast the Brown's model simulations of Middle Earth reveal the power of scientific prediction. These models can simulate not only the climates associated with past and future configurations of planet Earth, but also the climates of other planets and even fantasy worlds. Yet we must recognise the limited scope of such predictions. The physical climates of Middle Earth might be revealed through these models, but it takes Tolkien's imagination to bring to life the *significance* of these climates for the characteristics and destinies of the dwellers of Middle Earth. The ways in which humans, hobbits, elves and orcs become weathered through climate is beyond scientific prediction.

\*\*\*

The production of knowledge of climates as yet unknown, unseen or unrealised – knowledge described variously as forecasts, prognoses, scenarios, projections or predictions – is only possible through cultural practice. Yet climate models and the knowledge they generate are only one example, even if a conspicuous one, of how cultures work to make, stabilise and lend authority to predictions of future climate. There are different cultures of prediction which have emerged at

---

[1] For further details, see University of Bristol (2013) 'Scientists simulate the climate of Tolkien's Middle Earth', http://www.bristol.ac.uk/news/2013/10013.html (accessed 16 May 2016).

different times in the past and which are still in circulation today. Religious prophecy, popular folklore and statistical forecasting are other examples of such cultural practices. Exploring these practices is the focus of this chapter. I consider, first, and in the context of prophecy, several wider forms of predictive practices that have been applied to thinking about future climates, before focusing specifically on climate models and modelling as a generator of predictive knowledge. A third section reflects on the ways in which such model-generated predictions are used in different social and cultural settings.

## Prophets and Prophecies

There are many aspects of the future about which humans would like foreknowledge: their future health for example, or the safety of their children or the security of their job or livelihood. Perhaps because the atmosphere is so constantly in flux, and because weather has such a significant bearing on human life, it is not surprising that so many human cultures have sought to develop their arts of prediction with respect to the skies. And the precariousness of this task is also perhaps why these same cultures have so often developed ambivalent relationships with those who offer themselves publicly to construct and issue such predictive knowledge. Weather prophets have sometimes been valorised for the skills or status they possess to divine the weather. This may literally be 'divining', through discerning the will of the gods and spirits who are believed to control the weather. More mundanely, it may also be through their ability to observe closely the movements of the non-human world – the stars, clouds, birds or trees – which in some way disclose forthcoming weather.

Yet entering into the world of public weather prophecy or forecasting has also been a risky service to offer to society. Such knowledge may challenge existing epistemic or cultural norms and it may take only one disastrously wrong forecast to attract widespread ridicule. The originator of routine daily weather forecasts in the UK in the early 1860s, the meteorologist Robert FitzRoy, found this out to his cost. FitzRoy had gained fame earlier in his career as the captain of HMS Beagle, which carried Charles Darwin on his world-circumventing and world-changing five-year voyage. His qualities had also been recognised by the British Government through his appointment in 1842 as the Governor of New Zealand. Yet his later weather forecasting skills and claims, approximate at best, were castigated by *The Times* newspaper for transgressing the boundaries of what at the time was deemed to constitute credible and legitimate knowledge about the workings of the atmosphere. Despite his earlier feting by the establishment, his public reputation suffered and this exacted a toll on his mental health. FitzRoy committed suicide in April 1865 (Moore, 2015).

Similar questions and tensions about the legitimacy of forecasting also emerge with respect to predictions of future climate. On what epistemic basis can such predictions be made? Who has the cultural authority and public status to issue such predictions? What happens when predictions apparently turn out to be wrong? Yet predicting future climate brings additional difficulties relative to those of forecasting tomorrow's weather. The range of possible influences on the course of future climate increases considerably; it is not just the mood of the gods or the movements and cycles of the moon, sun and stars which might need to be considered, but also the state of the Earth's forests, oceans, ice-sheets and volcanoes, as well as the material consequences of human technologies and social practices. The spatial and temporal scales of the required insights are also challenged by the demands of foretelling how climates might evolve over much larger distances and for much further into the future than a forecast of the weather of a particular place might require. And the further into the future the climate forecaster applies her skills, so another challenge comes more sharply into view: the ambiguity of validation. If rain is forecast for tomorrow in Rome then the prediction can easily be validated and the weather prophet held to account. But if the next ice age is predicted to commence 100 years hence, the climate forecaster will long be dead before the claim can be validated. Other factors beyond empirical validation must usually be mobilised to build trust in the authority of the climate forecaster, as we shall see.

The cultures of climate prediction are many and varied. Noah discerned the voice of God to make his forecast of an extended global deluge, while the account in the biblical book of Genesis of Joseph in Egypt, and his prediction of alternating seven years of plenty and seven of famine, seems likely to have been based on his close reading of the hydrology of the river Nile, which reflected the oscillating climate of the Ethiopian headwaters. (Climatologist Stephen Schneider used this narrative as the inspiration for the title of his 1976 book on managing climate for human ends, *The Genesis Strategy: Climate and Global Survival*.)

An early example of climate prediction drawing upon the new sciences of modern Europe used statistical evidence of regular cycles in key meteorological variables such as temperature and precipitation. In 1890, the German geographer Eduard Brückner analysed many instrumental records to argue for a 35-year oscillation between cold-damp and warm-dry climates, which he linked to some unknown solar variation. On this basis, Brückner predicted a dry and warm climatic phase commencing early in the twentieth century and affecting American and Eurasia in particular (Brückner, 1890).

The growth in scientific understanding during the 1950s and 1960s of an interlinked Earth system and of the possible human influences on this system gave rise to speculation in the 1970s of Earth's climate either entering a new ice age or embarking on a sustained period of warming. Climate models were only at a very early stage of development and so an alternative option was pursued by the US Government to ascertain the climatic future: ask eminent climate scientists their opinion on the matter, a procedure

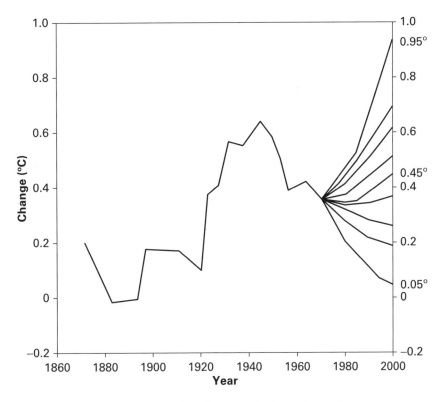

**Figure 9.1**   Expert projections of northern hemisphere climate (mean annual temperature change) to the year 2000, resulting from the expert elicitation exercise of the US National Defense University (Source: NDU, 1978).

known as 'expert elicitation'. In a 1978 study conducted by the National Defense University in Washington DC, the views of 24 experts drawn from seven different countries were solicited (NDU, 1978). They were each asked to judge the future course of northern hemisphere temperature out to the year 2000, with results relative to 1975 that ranged from a cooling of 0.35°C to a warming of 0.55°C (see **Figure 9.1**).

Expert elicitation as a method of prediction has a long history, and is traceable back to the Oracle of Delphi in Ancient Greece. Most societies have sought to establish 'oracles' of various sorts, experts to whom political leaders and society at large can turn in moments of crisis and uncertainty. Lynda Walsh outlines this fascinating history in her book *Scientists as Prophets: A Rhetorical Genealogy* (Walsh, 2013). She shows how in most contemporary societies the scientist or the science advisor has become this trusted source of prophetic knowledge; this is most certainly true at the present time with respect to climate prediction. And yet at the heart of this history is a struggle to understand the precise nature of the cultural authority such prophets are granted (see **Box 9.1**).

## Box 9.1: Climate Scientists as Prophets

At moments of uncertainty or crisis a society will turn to its prophets to search for guidance, for clarity about how the future will evolve and insight into how to act. This has been witnessed over the last half century with respect to the fate of global climate, the prophets in this case being climate scientists. The lack of clarity in the 1970s about the climatic future (see **Figure 9.1**) would eventually lead to the world's governments, through the United Nations, establishing the Intergovernmental Panel on Climate Change (IPCC) in 1988. The original brief of the Panel's experts was to assess all scientific evidence about the future of climate and its impacts, and to formulate appropriate response strategies.

By placing climate scientists in the cultural tradition of the prophet, Walsh helpfully untangles the confusions and conflicts which climate predictions and the reports of the IPCC have generated in contemporary societies (Walsh, 2013). At the heart of this confusion is the role and cultural expectation of the climate scientist. Is it their function to reveal the future as surely and accurately as possible, speaking from a stance of disinterestedness and objectivity? Or is their role less about a dispassionate unveiling of the future and more about entering into a public dialogue about what Walsh refers to as a society's 'covenant values', those values a polity shares and which distinguish it from its neighbours?

Adherence to the former expectation is philosophically attractive – it separates the 'is' from the troublesome 'ought' – but, Walsh argues, it is rhetorically unstable. As increasingly technical societies seem to rely increasingly on expert advice about the future consequences of technological activity, scientific advisors are caught in an ethical catch-22. For example, governments call upon the IPCC to predict the future climatic consequences of their actions. When these answers align with the existing rhetorical ambitions of these same governments, scientific advisors are lauded. But when critics want to challenge these policies, or when the climate predictions misalign with prevailing policy, the is/ought boundary is then invoked by politicians to undermine the ethos of the scientist. This manoeuvre was seen at work with respect to the Climategate affair and the subsequent challenges to the ethos of the IPCC during the winter of 2009/10. As Walsh concludes, 'Until some seemingly more certain civic epistemology evolves, we will seek our prophets among the sciences, and they will continue to engage us in a dialogue that goads us to recall our covenant values'. And following in the line of the prophetic tradition, '... although our science advisors cannot tell the future or tell us what to do, they do – and will continue to – help us to know ourselves' (Walsh, 2013: 198).

## Predictive Models

Central to the practice of climate prediction today is the climate model. Climate modelling is characterised by a distinctive culture of prediction, different to those mentioned in the previous section: religious knowledge, weather-lore, statistical extrapolation or expert elicitation. Taking inspiration from the new practices of numerical weather prediction in the 1950s and the revolutionary new capabilities of supercomputers, climate models emerged in the 1960s as powerful tools for organising scientific knowledge about the physical processes connecting land, atmosphere and ocean (see **Chapter 3**). Although the ability of the early generation of models to *simulate* Earth's climates was the incentive for this novel epistemic community of modellers, it was the promise of *predicting* future climates which drew increasing resources and patronage from governments through the 1970s and 1980s. Following the creation of the IPCC in 1988, climate modellers successfully fought off the challenge of palaeoclimatology as a rival epistemic community which also sought prediction of future 'warm-world' climates. In 1995, the first Climate Modelling Intercomparison Project was established under the auspices of the WMO, embracing more than a dozen climate modelling centres from around the world. This included the UK's Hadley Centre for Climate Prediction and Research, a national climate modelling capability created in 1990 through Prime Ministerial patronage (Mahony and Hulme, 2016).

Despite the spectacular ascendancy of model-based predictive knowledge of the climate system, or indeed because of it, the question remains: 'How do such models gain their cultural authority?' This is a question I looked at earlier in the context of other forms of climatic knowledge (see **Box 3.2**). Models are socially or culturally legitimated, as much as they are epistemically validated. Indeed, as Naomi Oreskes and colleagues argued many years ago, model confirmation is logically impossible; the primary value of climate models is heuristic (Oreskes et al., 1994). A model is not a 'real' thing; it is an organised abstraction of reality, designed to be useful. Understood this way, the characteristic cultural practices of modelling climate need not be limited to climate scientists (see **Box 9.2**).

---

### Box 9.2: Modelling as Cultural Process

The word 'model' is one of those heavily over-worked words in the English language. It can function as noun, adjective or verb. As noun, a model is a representation of an object, usually in miniature form, which reveals the appearance of something; for example, a clay figurine. As adjective, model is used to describe an ideal or exemplary type, for example, a 'model' husband. As verb, to model is to give shape or form to something, for example, to fashion something out of wax or to 'model' the climate using computers.

As an activity, modelling is something that can have different forms of cultural expression, not least with regard to climatic knowledge and prediction. Anthropologist Kirsten Hastrup has studied the practices by which people interpret the world around them to construct 'models' which enable predictions of

*(Continued)*

*(Continued)*

future climate. She finds much in common between the models of climate constructed by Earth system scientists and those constructed of their local climatic environment by Greenlandic seal-hunters. The climate models that result in each case are abstractions of reality, but are nevertheless capable of generating predictions which, in their different respective cultures, are deemed trustworthy and found useful; for example, predictions of future temperature extremes in the case of climate scientists or predictions of the distribution of sea-ice in the case of seal-hunters.

These very different modelling processes in fact rely upon five similar components: observation, formalisation, experimentation, projection and action. Observation is about paying careful attention to one's surroundings, whether or not aided by scientific instrumentation. Formalisation is about establishing rules and regularities in these observations, whether through mathematics or mental maps. Experimentation implies testing these rules through manipulating some form, whether computationally or experientially. Projection is the outcome of experimentation, predicting some new state beyond the reach of mere observation. And finally action, whether an every-day or political act which changes the world, is an outcome inspired by a projection. These modelling practices are marked throughout by cultural beliefs (respectively, the physicality of climate for scientists and the impor-tance of place for seal-hunters) and the power of symbols (respectively, for example, mathematics and ice shapes). Whether for scientist or seal-hunter, these five components form an interactive process for making reliable and trustworthy knowledge about possible future climates, or what Hastrup refers to as 'anticipations of nature'.

(Sourced from Hastrup, 2013)

One of the challenges for all those engaged in the prediction of climate, and one which applies acutely to climate modellers, is how to embrace and com-municate uncertainties about the future. When called upon to make public predictions of future climate – to be called upon for their prophetic voice (see **Box 9.1**) – modellers have multiple responsibilities to fulfil, the competing demands of which are not always easy to navigate or balance. On the one hand, as those who have been granted a certain cultural authority for their ability to 'see into' the climatic future, they need to demonstrate that their projections are both credible and useful. On the other hand, they also need to recognise the limitations of their art, particularly when called upon to predict future climates at sub-continental scales or far into the future (Hargreaves and Annan, 2014). Increases in a model's spatial resolution may increase the preci-sion of a climate prediction, but not necessarily its accuracy, and the validation conundrum, alluded to earlier, applies to predictions which extend more than a decade or two hence.

As nearly all those who have reflected on the nature of climate models have recognised, they are imperfect predictive tools even if they have their uses, a state of affairs captured in the aphorism 'all models are wrong, but some are useful'. For example, after a detailed study of the assessment culture of the IPCC, science studies scholar Jessica O'Reilly concluded that 'Projections of future climat[ic] change cannot be anything but guesswork – highly technical, expensive, educated guesswork – but always with an element of "gap-filling", prognostication or intuition' (O'Reilly, 2015: 123). How this 'guesswork' is undertaken and how it is communicated to public and political actors who may have high expectations of the truthfulness of climate model predictions is fraught with danger: danger for the credibility of science, for scientists' prophetic ethos and for political authority and policy-making. There are temptations for climate modellers to over-play their hand when communicating the results of their predictive simulations. How uncertainties and model limitations are communicated by scientists internally, to their peers, may well differ from how these same limitations are expressed in modellers' external utterances and perceived by wider audiences.

The so-called MacKenzie 'certainty trough' can be used to analyse these different perceptions of certainty which grow around climate modelling, or around what has been termed the culture of 'seductive simulation' (Lahsen, 2005). The metaphor of the certainty trough suggests that those who are closest to the processes of model development (scientists) and those who are furthest away (the general public) are likely to be more sceptical of the predictive claims of models than those who are professional users of model predictions for planning and decision-making (see **Figure 9.2**). These latter groups fall in the middle of the certainty trough, where scepticism is least. However, Myanna Lahsen's work amends this picture by suggesting that those

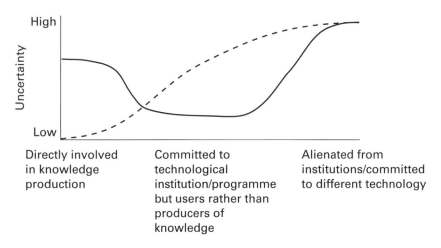

**Figure 9.2**   The revised 'certainty trough'. The dashed line indicates perceived uncertainty levels not accounted for in MacKenzie's original work (Source: Lahsen, 2005).

who have invested most in developing climate models, professionally and emotionally, might in fact also be blind to some of the limitations of their models for predictive purposes. As Lahsen concludes her study, '… modellers at times downplay model inaccuracies because they are interested in securing their [public] authority'. The shape of the MacKenzie trough should be adapted 'to account for limitations in modellers' ability to identify such inaccuracies' (Lahsen, 2005: 917).

## Using Climate Predictions

So do climate model predictions have practical value for a society? The aspiration to predict the climatic future, and the claim that climate models can partially fulfil that aspiration, has led some social actors and decision-makers to lean heavily on such predictions when designing climate adaptation strategies. The ambition here is to help societies 'weather' the climate, to assist people to live securely with their weather and its extremes. Reliance on climate model predictions has also been invoked to avoid maladaptation to possible future changes, adaptations which might perversely expose valuable people and places to greater climate risks. The human desire for certainty has given rise to the belief that climate predictions using scientific models will get ever more accurate. But this is a fallacy; it is the unfulfillable promise of prediction (Lemos and Rood, 2010). Uncertainty is intrinsic in scientific modelling and hence in climate prediction. It is questionable whether uncertainties embedded in climate predictions on time and space scales that matter for human decision-making will in any meaningful way narrow over time.

Two cultures of adaptation decision-making have therefore arisen in recent years, both of which draw upon climate predictions but in different ways. On the one hand is the 'predict-and-optimise' mode of decision-making. Climate predictions for decades or longer in the future are used to design infrastructures and management systems which will be 'weather-proof' on these timescales. The implication here is that more precise predictions leads to better adaptation. On the other hand is the 'robust decision-making' paradigm, which conceives predictive knowledge about climate differently. Rather than optimising against some predicted future, infrastructures and systems are designed to be robust against a wide range of possible future outcomes, no single one of which can be guaranteed. The implication here is that good adaptation is not conditioned on the accuracy of the prediction, which in any case cannot be inferred merely from the precision of the forecast.

This distinction in approach points to the value of Walsh's (2015) contrasting of different public expectations of the role of prophets. The uncomplicated view is that prophets – in our case climate scientists – are seers of the future, offering objective knowledge and (increasing) clarity about how events will turn out. Their role is to offer a firm basis for planning and decision-making. Conversely, prophets – historically understood and, Walsh argues, still today – are those voices in a society which, given a shadowy premonition or uncertain forecast of the future, can help that society make appropriate choices through reconnecting it to its covenant values (see **Box 9.1**). In this way, she argues, do cultures best manage the future.

With respect to future climate and adaptation decision-making this latter prophetic role might be a better one for climate advisors to envisage. The benefit of uncertain climate predictions for decision-making will always be conditioned by the range of indeterminables that condition the future and by a particular culture's experiences and values. Carla Roncoli and colleagues offer a good example of such precaution amongst rainfed farmers in dryland west Africa. Seasonal climate forecasts derived from scientific reasoning and models were issued a few weeks ahead of the rainy season by the meteorological office of Burkina Faso. How farmers used such predictive knowledge was heavily conditioned by their own observations, experiences, cultural practices and values. Climate predictions emanating from one epistemic culture – science – were reconfigured through the meanings and values of a different culture, that of the Burkinabé farmers. The meaning of these climate predictions was shaped by '… their cognitive landscape … [which] is constituted not only by the meteorological phenomena, but also by the accumulated experience of previous climate events and by conflicting hints, hopes and fears about the upcoming season' (Roncoli et al., 2003: 197).

Humility and precaution may well be the attributes that societies most need to cultivate with respect to the climatic future, over and above claims of accurate foresight and special knowledge. For societies to 'weather well' they might best be advised not to expect or demand ever greater certainty about what lies ahead from their prophets. A more judicious course of action might be to reflect on those adjustments and adaptations – both in infrastructural arrangements and in institutional cultures – which protect assets and social functions in the face of an uncertain climate and an open future. Climate may well have evolved as an idea which helps stabilise social relationships between weather and culture, yet when physical climates themselves are changing then the public deployment of the idea of climate needs reimagining. This is a reflection to which I return in **Chapter 12**.

## Chapter Summary

In this book I am arguing that climate is an idea which mediates between weather and culture. If this is so then foreknowledge of future physical climate has significance both for anticipations of tomorrow's weather and for the ways in which people live with this anticipation. Climate predictions may therefore hint that tomorrow's weather will be different from yesterday's weather, but such predictions – whether correct or not – also change the ways in which a society makes decisions today. As with all climatic knowledge, climatic predictions are culturally conditioned and how such knowledge is used varies from culture to culture. Different cultures of prediction yield different types of predictive claims and carry different degrees of public trust and authority. Predictions of future climate, whilst often relating in some way to scientific knowledge claims, are always mediated by a wider variety of cultural norms, computational artefacts and communicative practices. Numerical computer modelling of the climate is but one means of predicting its future course, but one that has taken centre stage in the contemporary world. Yet model predictions, and those who communicate them – standing in a long line of prophetic voices – face multiple challenges: how to be credible, how to carry authority, how to be useful. Navigating between these demands is not easy.

Climate prediction serves the human desire for certainty about the future, yet cannot escape the fact that such certainty is elusive. People therefore remain trapped between wanting to know the future and yet being unable to fulfil this desire. The usual state of affairs through human history has been one of 'not knowing' rather than of 'knowing' the future. The role of the climate prophet in such a situation is less about reducing this frustration through offering 'certain' knowledge of the climatic future. It is more about introducing a distinctive voice in society which provokes that society to find its *own* voice, interpreting and acting upon the irreducible uncertainty of the future through the covenant values it holds dear. And these values will vary from culture to culture.

It is therefore perhaps more helpful to interpret climate predictions, whether from scientific models, seal-hunters, religious divines or weather-lore, as stories about the future, 'cautionary tales that expose the dangers, taboos and prohibitions of certain courses of action' (Tyszczuk, 2014: 45). These stories of the future will affect the present, influencing cultural discourses and political actions. Portended climates may invoke dystopias and fears of loss (see **Chapter 7**), but they also can engender faith in human capability to bring about a 'better' future. Predictions will nearly always turn out to be wrong; the future has its own unfolding logic of action beyond human reach. But cautionary tales about the future can always be absorbed, pondered and usefully learned from. In the next chapter I investigate one mode of engagement with the climatic future where cautionary tales are particularly relevant: that of climate engineering.

## Further Reading

Anderson, K. (2005) *Predicting the Weather: Victorians and the Science of Meteorology.* Chicago, IL: Chicago University Press.

Hastrup, K. and Skrydstrup, M. (eds) (2013) *The Social Life of Climate Change Models: Anticipating Nature.* Abingdon: Routledge.

Moore, P. (2015) *The Weather Experiment: The Pioneers Who Sought to See the Future.* London: Chatto & Windus.

Walsh, L. (2013) *Scientists as Prophets: A Rhetorical Genealogy.* New York: Oxford University Press.

# 10

# Redesigning Climate

## Introduction

People have long sought to make their climate more agreeable, whether by emigrating in search of more favourable ones or by adapting locally to a climate's demanding vicissitudes. However, the idea of deliberately redesigning climate itself, beyond the micro-scale of forest clearances, walled gardens or painted roofs, is more recent. Schemes for modifying climate at macro-scales – whether to reverse a climatic deterioration or to enhance a climate's performance – can be linked to the modernist instinct for grand technological projects of environmental protection or improvement. This instinct emerged powerfully in European, and then later in American and Russian, cultures of the nineteenth and twentieth centuries, fuelled by the growth of science and engineering and encouraged by imperial projects of colonisation and superpower rivalry.

One such grand scheme for climatic improvement took shape in the 1920s in the mind of Herman Sörgel, a middle-aged Bavarian architect. Sörgel was troubled by a perceived decline in European culture, environment and economy, and by a weakening of Europe's geopolitical supremacy. He was not alone amongst Europeans in holding such a pessimistic outlook in the interwar years, but his solution was to think big. Sörgel proposed the creation of a joint Euro-African supercontinent to be invigorated through re-engineering the Mediterranean Sea and the river basins of Africa. His scheme would improve the climates of Europe and Africa and provide almost unlimited amounts of renewable hydroelectric power.

Sörgel's project was called Atlantropa and would commence with building an oceanic dam at Gibraltar to connect Europe and Africa and close off the Mediterranean Sea. Water levels in the Sea would fall through evaporation, creating a gradient of potential energy between the Atlantic and the Mediterranean water bodies. This energy would be captured through massive electricity generating turbines yielding 120 GW of electricity, six times the amount of electricity generated by the world's current largest power scheme at The Three Gorges dam in China. New land for settlement around the retreating Mediterranean coastline would be created and the climates of southern Europe and northern Africa would, he believed, be enhanced. Similar massive hydro-electric power schemes in the Chad, Congo and Zambezi Basins would achieve comparable benefits for Africa: vast renewable energy resources and the transformation of a climatically hostile environment – hostile, that was, for settling Europeans. During the 1930s Sörgel gained a small following of enthusiastic engineers and ambitious thinkers and he tried, unsuccessfully, to persuade Hitler's regime in Germany of the viability and benefits of his vision. Nothing came of the idea however, and by the time he died in 1952 Europe and the wider world were very different places to the way they had been 25 years earlier (see Lehmann, 2016).

Sörgel's vision of Atlantropa is instructive for thinking about the human desire to redesign climate through planned technological intervention. He was driven by a passion to harness the power of nature and by an irrepressible belief that human ingenuity could improve climates and enhance economic and social well-being. But we should recognise, too, how culturally situated Sörgel's ideas were. His vision was narrowly Eurocentric and marked by an extraordinary 'faith in large-scale planning, the trust in authoritarian rule, and a strict adherence to visions of imperial and racist domination' (Lehmann, 2016: 89). Atlantropa emerged at a time of perceived cultural and political decline in Europe, with concerns about limited fossilised energy reserves and an emerging awareness of the realities of climatic change. If climates were indeed changing, for whatever reason, then why not intervene to ensure that these changes were directed to some (humanly) desirable outcome? Concluding his study of Sörgel's ideas, the historian Philipp Lehmann observed,

> … with the new reach and scale of industrial engineering, what reason was there not to attempt to change climates on a continental, or even planetary, scale? With this approach to 'fixing' the global environment with the full force of modern technology, Sörgel also stands as one of the ideological forefathers of today's geoengineering schemes to halt and possibly reverse global warming. (Lehmann, 2016: 99–100)

\*\*\*

In this chapter I explore some of the cultural dimensions of the human desire to redesign climate. Although this ambition has found various expressions in recent centuries, visions and proposals have rarely, if ever, been converted into physical realities. Yet today there is a new tone and urgency to the latest set of ideas for climatic redesign. The possibility of global climate engineering – whether through removing carbon dioxide from the atmosphere or through altering the atmosphere's reflectivity – has gained considerable attention in some science, engineering and policy circles and engaged the creativities of various cultural entrepreneurs (contrast Keith, 2013, with Hulme, 2014a). Putative climate engineers, standing in Sörgel's tradition, are today proposing and researching technologies that would seek to 'manage' global temperature, much as the householder manages the comfort of her home or car by turning a thermostat.

The first section in the chapter explores the cultural origins of this desire for climatic perfection and control and how it has been expressed in schemes that seek to redesign climate. The following sections then elaborate two contrasting justificatory narratives for climatic redesign, one of climate improvement and enhancement and one of climate restoration and healing.

## Climates of Desire

Few people would consider their climate to be optimal – it is, for example, deemed too unpredictable (or else too predictable), too wet (or else too dry) or too cold (or else too hot). If climate is 'what people expect' and weather is 'what people get', then there will indeed be frustration with the gap between expectation and delivery. Although climate is an idea which promises to stabilise weather–culture relations (see **Chapter 1**), it never quite lives up to its promise. Expectations of what climate 'should' be like are always biased in the direction of human desire: preferences which arise in the imagination and are only loosely anchored in experiences of reality. It is for this reason that people in different eras and in different cultures have embarked on projects to imagine or re-create climates more desirable than the ones they have already (see **Box 10.1**). For example, in one study of utopian novels the authors found that this particular literary genre usually presents climate 'either as an equable given or [else as] something totally under man's [*sic*] control' (Porter and Lukerman, 1976). In the western imagination at least, the climate of 'paradise' – of an untrammeled and perfect Eden – is a climate of blue skies, tropical breezes and endless sunshine, an imaginary that continues to be used in marketing tourist destinations to affluent extra-tropical travellers. Visual depictions of the mythical Garden of Eden never convey rain or even threatening cloud; yet the vegetation is always green and lush, the grass never dry or parched. Eden's idyllic climate is tirelessly subtropical, a fantasy of utter meteorological stasis and balance.

---

### Box 10.1: Desirable Climates

People have found many ways to 'improve' their climate. Our distant hominin ancestors, so we are led to believe, walked out of Africa in search of more favourable climates. They successively colonised and abandoned land masses further north and east as climates waxed and waned through the Quaternary era. Writers in antiquity frequently valorised certain climates, but only after

*(Continued)*

*(Continued)*

they had been discursively purged of their most disagreeable components – storms, blizzards and leaden skies. These optimised imaginary climates then happily coincided with the temperate climates of the Classical world, well away from torrid equatorial and frigid polar zones. And medieval Europeans were able to articulate their own Edenic climates, as in this description from the fifteenth century of the imaginary Land of Cockaigne: 'There is no heat or cold, water or fire, wind or rain, snow or lightning, thunder or hail. Neither are there storms. Rather, there is eternally fine, clear weather… It is always a wonderfully agreeable May' (Pleij, 2001: 180–1).

By the eighteenth century, European colonisation and new technologies for land improvement were offering the prospect that at least some utopian climates might become reality. The French philosopher Comte de Buffon, recognising humanity's growing imprint on the physical forms and processes of the planet, could write that 'Man will be able to alter the influence of its own climate, thus setting the temperature that suits it best' (de Buffon, 1778: 244). Draining swamps and clearing forests for agricultural development were seen as part of a project of climatic improvement to which American settlers and European colonists of the tropics believed they were called. But not just in the tropics. The French socialist visionary Charles Fourier was concerned about the deteriorating climate of the 1840s. Amongst the utopian projects of his American disciples was 'the regulation of the seasons, the moderation of temperatures, and the control of climates, in such a way as to have them always the most favourable' (cited in Meyer, 2002: 592). Fourier's global atmosphere was to be serene and genial, rather like the eternally 'agreeable May' of the Land of Cockaigne. And then 100 years later, the nuclear scientist and mathematician John von Neumann foresaw the prospective power of climate control being realised through accelerating technological prowess, converting utopian dreamings into climatic realities. Von Neumann wrote that '… intervention in atmospheric and climatic matters will come in a few decades and will unfold on a scale difficult to imagine at present … what power over our environment, over all nature, is implied!' (von Neumann, 1955: 41).

These past efforts to 'perfect' climate – whether through emigration, behavioural adjustments or technological engineering, or simply through dreaming – all point to a latent dissatisfaction with the existing climate, or at least to a fear of losing a climate which is already known.

(Sourced from Hulme, 2014a)

Modern and western sensibilities with regard to what constitutes a desirable climate are just that: modern and western. What is 'optimal' climatically speaking has varied from place to place and across different historical eras. As I showed in **Chapter 2**, at the beginning of the twentieth century Ellsworth Huntington and geographers of similar ilk found empirical 'proof' that the optimal climates for economic wealth production were those of southern England and the north-eastern seaboard of America. On different timescales climate historians label the milder and wetter global climate of the early Holocene era, between 5,000 and 8,000 years ago, as 'optimal', implying that more recent climates have been sub-optimal. For example, Brian Fagan titled his 2004 book about the early Holocene *The Long Summer: How Climate Changed Civilisation*, implying the desirability of these early Holocene climates. Historians also describe the climates of the North Atlantic region between the eleventh and fourteenth centuries as representing a secondary (Medieval) 'climatic optimum', after which European climates deteriorated. Thus Geoffrey Parker's 2013 account of human society in the less 'optimal' conditions of the seventeenth century is titled *Global Crisis: War, Climate Change and Catastrophe*. Conceptions of what constitutes a 'desirable climate' are far from stable, as I showed earlier with regard to the Caribbean (see **Box 2.2**). The climates of these sub-tropical islands, formally regarded by Europeans as unhealthy and dangerous, were re-imagined in the twentieth century as desirable winter destinations for northern travellers. Hence a 1921 government brochure could boast that Barbados offered 'the healthiest spot on the globe' and promised to be 'the sanatorium of the West Indies' (Carey, 2011: 140).

It is the power of these complex cultural imaginaries that has fed the human appetite for schemes of climatic control and engineering. On the one hand these schemes can be designed to protect what is already desirable or to prevent a climatic deterioration to some less desirable condition. On the other hand, as in the case of Sörgel's Atlantropa, they may be intended to enhance climate by overcoming the poor climatic hand which has apparently been dealt to some regions. It is this human restlessness and frustration with climate which has nurtured dreams of climatic redesign. These have ranged from Charles Fourier's socialist blueprint for a human-engineered climate (**Box 10.1**) to the adoption of climate control as a weapon of war, as imagined during the Cold War decades of the 1950s and 1960s. For example, in a *Newsweek* article from 1958, the columnist wrote, 'The question is no longer, "Can man modify the weather and control the climate?" but "Which nation will do it first, the United States or the Soviet Union?"' (Anon, 1958). And so the new climate dreamers of the twenty-first century, environmental engineers such as David Keith and James Lovelock, who propose to geo-engineer the atmosphere and oceans, inherit a long human tradition of seeking to manipulate the world's climate.

These projects of climatic redesign can be interpreted through very different metaphors of human agency. Environmental theologian and ethicist Willis Jenkins has argued that it is important to pay careful attention not simply to metaphors that are used to describe *nature* (Jenkins, 2005). Scrutiny is equally important of metaphors that are used to describe *human agency* with regard to the natural world. He offered two guiding criteria for evaluating metaphors of agency. Well-chosen metaphors should, first, capture the complexity of there being various degrees of 'the natural' and 'the artificial'. Second, they should accommodate a productive role for human interventions whilst still being alert to the degenerative potential of such actions. Thus 'perfecting' or 'caring for' nature would, he argued, win out over 'preserving' or 'managing'.

So it is worth asking what metaphors of human action could be used when reflecting on schemes to intentionally redesign the Earth's climate? Current proposals for climate engineering intervene in the physical dynamics of the climate system through either removing greenhouse gases directly from the atmosphere or else deliberately changing the reflectivity of clouds or stratosphere. How should the intentions and goals of these interventions be described? What stories are being told and what is the self-perception of the people who seek to implement them? The following sections reflect on two very different narratives, or metaphors, of climatic redesign: one of climatic improvement and enhancement and one of climate restoration and healing. These justificatory stories of climatic intervention draw upon radically different philosophies of nature and accounts of human agency and intention.

## 'Improving' the Climate

The metaphor of improvement (and its associate enhancement) is certainly one that would appear appropriate to describe the motivation behind many proposed projects of climatic redesign. As we have seen, this was explicitly the way in which Sörgel imagined his Atlantropa project: the improvement of Africa's climate for European settlement and the enhancement of southern Europe's climate to better serve human needs. Such metaphors also apply to many of the grandiose and sometimes crackpot ideas for climate modification which emerged in nineteenth-century America, a catalogue of comedies recounted by Jim Fleming in *Fixing the Skies: the Checkered History of Weather and Climate Control* (Fleming, 2010). In an editorial in the *New York Times* from 1891, some of these ambitions were satirically lampooned. Schemes for climatic improvement that had been proposed by American entrepreneurs included unseasonable frost elimination, the dissipation of snow-storms and rain-making through inducing mid-air explosions or 'airquakes'. The *Times'* editorialist remarked, sarcastically, 'Should [these schemes] accomplish all that they promise, our liberal and enlightened Government will, doubtless, soon be able to relieve nature of a great part of her burdens, buy her out at a handsome price and run this section of the universe according to its own notions of what is right and proper' (Anon, 1891).

Although Atlantropa died with Herman Sörgel in 1952, the new politics of the Cold War and the rise of massive state-sponsored investments in science, technology and engineering by both Soviets and Americans alike, led to a new surge of ideas for large-scale climate engineering in the 1950s and 1960s. Some of these suggestions – few of them ever reached the stage of serious proposals – are pictured in **Figure 10.1**, reproduced from a thoughtful article on the topic published in 1974 in the leading journal *Science* (Kellogg and Schneider, 1974).

As with Fourier in post-Revolutionary France, climate modifications in nineteenth-century capitalist America, or Sörgel in culturally depressed interwar Europe, ideas for climatic enhancement should always be understood, indeed are only imaginable, within particular cultures. In the case of the schemes depicted in **Figure 10.1**, these postwar cultures were shaped by superpower rivalry, by the still ascendant faith in human technological prowess, and by ideological narratives of control over nature. Writing half a century after these schemes were suggested, the American sociologist John Lie explicates the cultural myth upon which the desire to improve climate for human benefit rested: 'Even if God-like control is unattainable – and the very idea of the conquest of nature is hubris – a deeply optimistic strain in the scientific mindset envisions nature as increasingly colonised and controlled by human conceptual inventions and technological interventions' (Lie, 2007: 234).

As Kellogg and Schneider commented in their 1974 article 'Climate stabilisation: for better or for worse?' these various suggestions for climatic enhancement encountered serious practical, ethical and political objections. What would be the consequences of the schemes proposed for other aspects of the climate system? Who would decide what should be accepted as the 'optimum climate' outcome to which these proposals were directed? And given that any scheme of climatic improvement would have both winners and losers, how would losers be compensated and by whom? The authors perceptively observed, 'We have the impression that more schemes will be proposed for climate control than for the control of the climate controllers' (Kellogg and Schneider, 1974: 1171).

More than 40 years later these questions of ethics, governance and justice remain live and pertinent with respect to the latest set of proposals for global climate modification, even if now argued less as technologies for climate improvement than as (emergency) interventions to arrest further climatic deterioration (see **Box 7.2**). These proposals include stratospheric aerosol injection, marine cloud brightening, orbital mirrors, urban whitewashing, soil biochar[1], ocean fertilisation, direct air capture of carbon dioxide, enhanced weathering of carbonate and silicate rocks, and more besides. The seriousness with which some of these suggestions are being taken is reflected in the growing range and number of specialists and public commentators who have engaged with the ideas. Scientists, philosophers, ethicists, theologians, political scientists, economists, artists and other cultural entrepreneurs are all entering into public conversations about the desirability or not of such ventures. In this sense, and in others too, there seem to be interesting and important parallels between the ambitions of climate engineering and those of the transhumanist movement (see **Box 10.2**). One seeks enhanced biophysical technologies to secure the planet from the damages caused by fossil-fuelled industrial capitalism; the other seeks enhanced biomedical technologies to liberate humans from the physical constraints of their evolutionary past.

---

[1] Biochar is a solid material obtained from the carbonisation of biomass and which may be added to the soil to improve its properties. See: http://www.biochar-international.org/ (accessed 11 June 2016).

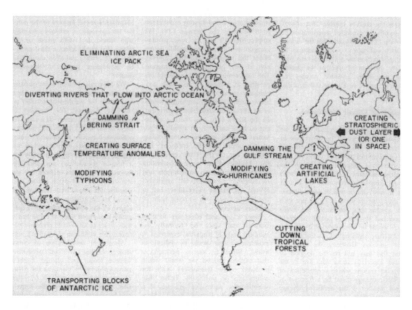

**Figure 10.1**   Schematic illustration of engineering schemes that had been or could be suggested to improve or enhance the climate (Source: Kellogg and Schneider, 1974).

---

## Box 10.2: Climate Enhancement

Developments in biophysical and neurological science in recent decades have enabled intervention technologies which in turn have contributed to the idea of human enhancement, the making of 'better' human beings. This narrative underpins the wider transhumanist movement, which seeks to enhance human intellectual, physical and psychological capacities, leading to the possibility, it is claimed, of post-human beings. The leading transhumanist Max More comments: 'With all due respect, we must say that you [Mother Nature] have in many ways done a poor job with the human constitution... We have decided that it is time to amend [it]' (More, 2013: 449). There are interesting parallels here with the narrative of climate improvement, redesigning the atmosphere to correct the faults or deficiencies introduced by this 'poor' and imperfect human specimen.

As with the human body, the physical climate has been and continues to be changed inadvertently through a wide range of socio-technical practices rooted in particular human behaviours. And so in both domains – the body and the climate – projects of enhancement, correction and improvement have emerged. As human enhancement technologies multiply, applied to enhancing mind, memory, beauty and physical strength, so too are calls emerging for the creation of wilfully 'corrected' and enhanced weather. A climatic control technology such as stratospheric aerosol injection exemplifies such aspirations (see Keith, 2013).

The philosophical, moral, ethical and political questions that apply to human enhancement apply equally to the prospect of climatic enhancement. Both weather and human behaviour remain significantly unpredicted, if not unpredictable and this limitation persists under conditions of climatic or bodily enhancement. So how benign for the weather/the body are such technological interventions? How can one distinguish between the enhanced and the unimproved condition? How far do people desire to become masters of the skies/themselves? Who governs these technologies and for whose benefit are they governed? Michael Hauskeller has explored these questions with respect to human enhancement. Following the moral philosopher Michael Sandel, Hauskeller's apologia for being cautious, if not sceptical, about such enhancement is based on the importance of retaining a certain given-ness or giftedness about the human body (Hauskeller, 2013). This reasoning might also apply to the climate: it is not morally wrong to enhance the climate, but it may be unwise. Humans might create desirable climates, but in the process they might become undesirable people (Jenkins, 2016).

It is not only the imaginaries, narratives and schemes for climatic improvement that arise through the cultural imagination; it is also the possible consequences of such schemes going wrong that are given substantive form and visibility. Two recent films have developed such scenarios as the basis of their storylines, *The Colony* (2013) and *Snowpiercer* (2013). Both films begin and end in an ice age, but the dramatic changes in climate were caused by the malfunctioning of well-meaning technological efforts to restrain a warming planet. In *The Colony*, huge cloud-making installations were intended to reduce the sunlight reaching the Earth's surface, but so great was their output that the clouds obscured the sun and the planet froze. The few survivors end up living in underground caverns. In *Snowpiercer*, the spraying of sun-blocking aerosols into the upper atmosphere misguidedly produces an ice age. The remaining humans, incarcerated on a massive train which repeatedly circumnavigates the world, contend for survival amidst increasingly violent displays of class inequality.

Both films offer a commentary on the hubris of global climatic improvement which in each case leads to the near extinction of humanity. Given that unrestrained global warming, which these interventions were intended to arrest, was presumed to present humanity with an existential risk, these are indeed ironic outcomes. Yet the two films end differently. In *The Colony*, a heroic band of survivors, headed by the colony's African-American leader, resists gangs of nihilistic marauders before eventually locating a still-functioning cloud machine. They are able to reverse the process and re-regulate an engineered planetary thermostat. In *Snowpiercer*, after 17 years of circumnavigation of the Earth, class conflict finally derails the train and just two children (of colour) emerge from the wreckage – into a world which is brightly lit and warming, having 'healed' itself of the impositions wrought by human design. These films therefore display different, but related, imaginaries at work with regard to the enhancement of the world's climate and to the desire to participate actively in such improvement. And by placing people of colour as 'the saviours' and survivors of the wreckage of an engineered climate, both films seem to offer a comment on where in society, ethnically speaking, the hubris of global climatic enhancement lies.

## 'Restoring' the Climate

The language of restoration offers a very different metaphorical vocabulary to that of improvement or enhancement through which to think about the motivations of intentional climatic redesign. The intervention technologies being imagined may in each case be largely, or even entirely the same, but the justificatory narratives for climatic redesign disclose very different accounts of human agency and ambition. Nevertheless, as with climatic enhancement, the metaphor of restoration raises challenging questions about how humans become weathered through climate.

The metaphor of restoration is one that has been applied to conservation and ecosystem management for some time (Marris, 2011). In the face of the damage or destruction caused to ecological functioning by human activities over many generations, ecological restoration seeks to allow ecosystems to 'recover' their health, integrity and sustainability. Rather than seeking to improve or enhance climate – to make it 'better' – interventions described as restoring climate are claiming the ability to reverse-engineer the damage to the atmosphere wrought by human actions. In conservation biology, restoration is also closely related to the idea of re-wilding, re-introducing apex predators and keystone species into regions and ecosystems from where they had been made extinct by human actions. One might therefore ask whether schemes for climatic restoration are also about recovering something of the 'wildness' or 'naturalness' of climate.

The imaginary of restoration is also related to metaphors of cleansing, purifying or healing the atmosphere. For example, climate engineering technologies such as ocean iron fertilisation, direct air capture or enhanced chemical weathering are all focused on removing a substance – carbon dioxide – from the atmosphere, thereby restoring it to some pre-human (or at least a less-human) condition. It is not surprising therefore that one of the most frequently adopted metaphors in media representations of new climate engineering technologies has been that of the planet as 'patient/addict' (Nerlich and Jaspal, 2012) and climate engineers as 'climate physicians' (Fleming, 2014). If the world's climate has an unnaturally elevated temperature – i.e. a fever – then what the climate needs is the attention of the world's most skilful doctors – i.e. climate scientists (see **Figure 10.2**). For example, the President of the UK's Royal Society, Sir Paul Nurse, remarked in 2011 that 'Geoengineering research can be considered analogous to pharmaceutical research' (quoted in Nerlich and Jaspal, 2012: 143).

**Figure 10.2**   The medicalisation of global climate. Following the meeting in Madrid in 1995 to finalise the IPCC's Second Assessment Report, scientists Bert Bolin, John Houghton and Luiz Gylvan Meira Filho are caricatured evaluating the feverish condition of the Earth's climate (Source: *Nature* 455 (2008)[2], reproduced in Fleming, 2014).
© Springer Nature

For some, the 'restorative discourse' around climate redesign is appealing. Seeking to restore climate to some pre-industrial condition evokes Edenic and bucolic myths of nature 'purified' of human contamination. These myths have long and powerful residual appeal in many western and aboriginal cultures. And the associated restorative metaphors of healing, medicine and physicians trigger comforting pictures of beneficent, skilful and ethically upstanding professionals doing their best for a sick patient. Yet these appeals applied through different cultural registers can also generate unhelpful outcomes. The restoration metaphor perpetuates the notion that there can be a 'natural' world which is free from its co-evolution with humans and their multifarious activities. And the medical metaphor of planet as a patient/addict requiring healing can also trigger the emergency frame, whereby restorative medicine – climate engineering – is needed urgently, now, to avoid further illness. Thus emergency discourses carry serious dangers, opening the path for overly hasty or anti-democratic measures (see **Box 7.2**). Healing or restorative metaphors can easily be subverted by saying that climate engineering is like plastic surgery for the planet. Or as one commentator put it: 'Geoengineering is not a cure. At best, it's a Band-Aid or tourniquet; at worst, it could be a self-inflicted wound'[3]. Different metaphors of climatic redesign carry connotative effects which are not easily predicted or contained.

---

[2] http://www.nature.com/nature/journal/v455/n7214/full/455737a.html (accessed 17 May 2016).
[3] Conway, E. (2014) 'Just 5 questions: hacking the planet' NASA: Global climate change, http://climate.nasa.gov/news/1066/ (accessed 11 June 2016).

## Chapter Summary

People have long expressed the desire to 'improve' their climates, to make climate more congenial to their preferred types and sequences of daily weather. These desires have historically been pursued through migration, through micro-scale practices of technological innovation and management or through the exercise of the imagination. The changes that humans have now wrought on the global atmosphere have, however, introduced a new rationale for the intentional redesign of climate – the sense that unfettered human activities are leading to a less than hospitable climatic environment for human well-being and flourishing. Yet all projects of climate engineering are culturally situated, as I have illustrated with the case of Herman Sörgel and Atlantropa. The global climate engineering technologies now being discussed and evaluated (*in silico*) engage the human imagination in very different ways. For some, they are 'a horrifying idea whose time has come'; for others they are technologies which bring restorative or healing properties to the planet's climate, like medicinal drugs or surgical procedures for the body.

Climatic redesign can be imagined as offering an improvement or enhancement to climate, correcting either nature's deficiencies or the adverse consequences of human activities by taking control of climate and creating something better; in which case such technologies have parallels with transhumanism and the narrative of human enhancement. On the other hand, climatic redesign can also be conceived as a project which brings about the restoration of a more natural, less artificial climate. Climate engineering is then about recovering the imagined climatic stability of the Holocene, allowing the planet's climate to heal itself following the damage wrought by industrial capitalism. For example, this is the self-declared goal of the campaigning movement *350.org*[4], returning the atmospheric concentration of carbon dioxide to its pre-industrial level.

These different accounts of the purposes of climatic redesign appeal to different instincts in different people across many different cultural traditions. Perfected climates, it seems, can only be found by mastering climate on the one hand or by expunging humans on the other. But however conceived, such climate engineering practices raise profound philosophical and political questions. What would it be like for humans to live in an enhanced or restored climate, knowing that it was made and maintained by human hands? And, politically, who gets to decide and govern the climates thus designed – those who have most to gain, those who have least to lose or those who simply have the power and means to enact the technology? In the next chapter I consider some of the questions associated with the governing of climate.

---

[4] http://www.350.org (accessed 18 May 2016).

## Further Reading

Fleming, J.R. (2010) *Fixing the Sky: The Checkered History of Weather and Climate Control.* New York: Columbia University Press.

Hulme, M. (2014) *Can Science Fix Climate Change? A Case Against Climate Engineering.* Cambridge: Polity.

Keith, D. (2013) *A Case For Climate Engineering.* Cambridge, MA: MIT Press.

Morton, O. (2015) *The Planet Remade: How Geoengineering Could Change the World. The Challenge of Imagining Deliberate Climate Change.* London: Granta.

# 11

# Governing Climate

## Introduction

On 12 December 2015, the world's governments put their names to the newly negotiated Paris Agreement on Climate Change. This Agreement represented the most inclusive and ambitious instrument of climate governance yet constructed and was hailed at the time as a great triumph of international diplomacy. Its 29 Articles were designed to 'hold the increase in the global average temperature to well below 2°C above pre-industrial levels' and to promote further efforts to limit the increase to no more than 1.5°C. The Agreement recognised the importance of all levels of government – national, regional, local – and also the roles of non-governmental political and social actors for securing this goal. The document was founded on several principles, most importantly that of 'common but differentiated responsibilities and respective capabilities, in the light of different national circumstances'. Levels of international financing adequate for enabling climate-friendly development and promoting clean energy technologies were also stipulated.

The Paris Agreement was the most recent outcome of an international negotiating process which dates back to the signing of the United Nations Framework Convention on Climate Change (UNFCCC) at Rio de Janeiro in 1992. It followed the earlier 1997 Kyoto Protocol, the 2001 Marrakech Accords and the 2013 Warsaw International Mechanism for Loss and Damage, other international instruments of climate governance negotiated multi-laterally through the UN. This negotiating process has normalised the idea that global climate not only *can* be governed, but that it *must* be governed to avoid undesirable climatic outcomes. A cardinal principle of the process has been that the UN should orchestrate these efforts and that the overarching goal of climate governance is the regulation of global-average surface air temperature.

This idea of governing climate – not just that *global* climate can and must be governed, but even the idea that climate at *any* scale can be governed – is a relatively recent one. Yes, there is a long history of people exerting control over their local physical environments through deliberate and, at times, coordinated actions. Indeed, human social evolution of the last 10,000 years could be written as a story of an increasing ambition and realisation of societies to govern, or 'steer', the physical and social properties of their immediate environments. Yet for most of human history, governing climate has been as much about controlling the self or managing one's property well (see Meyer, 2002), as it has been about governing some external physical entity. Even in the modern era, until very recently, it has at most been about developing local or national techniques of environmental control.

If climate first emerged *as an idea* that helped people stabilise their cultural relationships with weather in particular places, as I have argued in this book, then how did it become possible for climate *as an object* to be deemed governable through international treaties and instruments? How did climate, not least global climate, come to be thought of as an entity that could be deliberately steered or directed in certain ways? Adopting Michel Foucault's idea of governmentality (Bulkeley and Stripple, 2015), at least two steps would seem necessary. First, climate needed to be 'objectified', to be turned into an external object of knowledge through authoritative and transferable epistemic practices. The emergence of science in the eighteenth and nineteenth centuries as a mode of knowing climate was important here (see **Chapter 3**). Second, this objectified climate needed to be 'problematised', to be recognised discursively as an entity which can exist in multiple physical states each of which carries different significance for different social actors.

As with any other physical entity, social category or cultural activity – for example, rivers, gender or sport – climate becomes political as soon as it is successfully objectified and problematised by one or more social actors or groups. For climate to be governable, it must of necessity become political. And for climate to be made political – to become an object of contestation – it becomes governable in principle. It therefore follows that to understand how climate is governed is not just to describe and explain the institutions and instruments of governance: such as the UNFCCC, the Paris Agreement, carbon taxes or emissions trading systems. It is also at the same time to reveal the interests of different social groups vested in the condition of climate and to disclose their differential access to, and demonstrations of, power.

It is not surprising that as climate has become an object of governance, it has become a battleground over which different political visions of the 'well-ordered society' or the 'good life' fight for supremacy. Climate has become a political object around which different modes of governing, regulating and ordering society vie for recognition, ascendancy and legitimation. Climate is, as one definition of politics puts it, 'an object of contestation' (Barry, 2001: 6). The imagination or prediction of possible future climatic conditions (**Chapter 9**) or projects to redesign climate (**Chapter 10**) are deeply political acts, bound up with the construction of knowledge, the development of technology and with demonstrations of power. Just as the *idea* of climate means different things to different people in different places, so the ambition to *govern* climate becomes a platform for the expression of different political visions. Governing climate entrains the human imagination, mobilises science and technology, and deploys various social institutions. But in the end governing climate is a political project: it is about who governs, how they govern, for what purpose and in whose interests.

***

In this chapter I explore different ways in which climate in recent centuries has been approached as an object of governance and some of the political implications of these governance projects. I do so by considering three different scales of governance ambition: local, colonial and global climate.

## Governing Local Climates

How climate is governed is inseparable from how climate is known, in particular from how different accounts of climatic change and human agency are written into this knowledge (see **Chapter 4**). If climate is thought of as immutable or beyond direct human influence then the idea of deliberately 'steering' climate makes no sense. Climate remains 'the domain of the gods' (Donner, 2007), but in this case gods who are impassive to human supplication. On the other hand, if physical climates are thought to be the outcomes of dynamic local processes of environmental change in which humans are implicated (even partly), then different possibilities emerge. For example, European and North American elites of the seventeenth to nineteenth centuries came to believe that local climates could be influenced by human actions – draining marshes, clearing (or planting) forests, cultivating land. Climate became objectified through the new scientific instrumentation of the eighteenth century and became problematised through the climatic theories of political philosophers such as Hume, Bodin and Montesquieu. The condition of climate mattered for upholding the social order, driving economies and affecting public health. In these centuries, new philosophical conceptions of the relationship between nature, climate and human agency opened spaces in which to imagine different ways that human activities could influence local climates. Taken together, these conditions facilitated the possibility of governing climate, at least local climate.

The early modern period therefore saw entrepreneurs and politicians begin to deliberate how local climates might be regulated for economic or social benefit. The precise means of governing climate would come to reflect the knowledge and politics of the day and the vested interests of the powerful. In pre-Revolutionary France for example, projects of land drainage served the goals of both civilisational and climatic improvement and were implemented with the support of the monarchy. Bringing local climates under political control in this way demonstrated the power and beneficence of the French state. Similar projects of local climate governance emerged in post-Revolutionary America in the mid-nineteenth century. The origin of the 1850 Swamp Land Acts, for example, lay in concerns which had emerged a generation or two earlier about the adverse climatic environments of wetlands and marshes in eastern coastal states. These climates incubated mosquitoes, the cause of devastating yellow fever epidemics in the late eighteenth century. Land drainage therefore became a means of improving local climates and reducing the prevalence of deadly disease outbreaks.

Governing local climate was not only an exercise in bringing forests or swamps under political oversight. For example, in 1806 the French doctor Eusèbe de Salverte surveyed the different climates and peoples of the Napoleonic Empire with a view to fitting populations to the most appropriate climates, governing the movement of people 'through' climate one might say. The French Government's new-found authority could also be used in 'major works projects ... to improve the "physical constitution of the climate" and – once again – that of populations' (Locher and Fressoz, 2012: 583). And as I showed in **Chapter 4**, Charles Fourier in the 1820s argued that the root cause of the deterioration of climate was social. While deforestation and land degradation might be proximate causes of climatic change, the ultimate causes were greed and individualism. Prefiguring more recent justifications for modes of climate governance which seek to discipline human behaviour, Fourier argued that climate could only be governed by governing the people, by transforming an individualistic society into a collectivist, socialist one. Forestry legislation, he said, was just a band-aid for a climate in decline; 'the only solution to planetary ills was revolution: "We need to get away from civilisation"' (quoted in Locher and Fressoz, 2012: 587).

Local climate governance has continued to emerge in new sites and to new ends. One recent example concerns the challenges of governing city climates. As cities have grown in size and intensity through the last century, so their urban form and socio-economic metabolism have contributed to the emergence of significant urban heat islands. City temperatures can be up to 3°C warmer on average than surrounding regions, on some nights up to 10°C warmer. The new science of urban climatology has objectivised urban climates over the last 50 years and the desire to regulate city climates for the benefit of urban populations, environments and economies has effectively problematised them. Tokyo offers one example of how far the governing of urban climates reaches into diverse realms of technical, social, ecological and behavioural control. The Japanese Government's 'Policy Framework to Reduce Urban Heat Island Effects' published in 2004 called for legislation to alter street surface materials, introduce street-tree planting schemes, design cold-air corridors, improve urban structure and modify human behaviour through, for example, the relaxation of office dress codes (Janković, 2015).

These examples of projects of local climate governance show the variety of interventions which have sought to bring climate under human control, thereby protecting climate as an idea that stabilises relationships between weather and culture. Local climate governance manifests in a variety of projects: transforming nature for human benefit (but also denouncing environmental regressions), segregating races, ruling populations, designing urban spaces. As I will emphasise later, governing climate is always about governing other things: social relations, mobility, energy, trade, investment, human aspiration, trees (e.g. see **Box 11.1**). This is as true for local climates as it is for colonial climates or global climates, to which I now turn.

---

### Box 11.1: Governing Climate through Forest Conservation

Understanding the relationship between forest dynamics and (local) climate has been a recurring quest over two millennia or more (see **Chapter 4**). It is not surprising then that governing 'the forest' has frequently been seen as a means of governing 'the climate'. Anxieties that deforestation was damaging to climate were particularly acute in the eighteenth and nineteenth centuries in, *inter alia*, North America, Europe, India, British and French island colonies, and New Zealand. Parliamentary or expert committees were established to consider what should be done in Prussia, Italy, Britain, Russia and New Zealand. For example, Brückner's concern about European climatic change (see **Chapter 9**) led him to conclude that climate needed protection through the implementation of forest conservation laws. He became instrumental in persuading the Prussian house of representatives to enact legislation to preserve the region's forests for the purpose of protecting regional rainfall and river levels (von Storch and Stehr, 2006). In light of increasing deforestation in the 1860s, similar climatic anxieties emerged in New Zealand. Unregulated felling was deemed to reduce rainfall and increase temperature and there were calls for felling to be reversed through state forest management. Advocates argued that government should increase its role in society and establish forest conservation laws to 'govern' the nation's climate (Beattie, 2009).

Regulating forests has also become a means of governing climate in more recent policy debates. However, the scale of intervention has moved from governing local and regional climates in the nineteenth century to today's goal of governing global climate. Forest degradation and clearance contributes a net increase of carbon dioxide in the global atmosphere, thereby warming the planet; about 30 per cent of all anthropogenic carbon dioxide in today's atmosphere originated from deforestation activities. The text of the Kyoto Protocol in 1997 first introduced the idea of tropical forest management as an instrument for climate regulation under the UNFCCC. The proposal eventually came into formal existence in 2007 as REDD (Reducing Emissions from Deforestation and Degradation) and was further refined as REDD+ in 2013. REDD+ is now established as an instrument of international climate governance which uses market mechanisms to incentivise national governments and local communities to 'protect' the carbon sink of their forests. As with many other instruments of climate governance, REDD+ raises challenging political questions, in this case about land tenure, theories of valuation and social justice. It illustrates one of the central insights contributed by the idea of governmentality: governing climate is not merely a technical or accounting issue; it is a political project which carries far-reaching implications for both the powerful and the powerless.

# Governing Colonial Climates

Climates became objects of governance in the European colonies of the eighteenth to mid-twentieth centuries. These ventures into climate governance were shaped by how (mostly) tropical climates were known or imagined by European cultural elites in the temperate north and by climatic theories exported from the metropolitan centres of political and scientific power. These theories understood climates to be intimately bound up with moral, economic and social well-being (see **Chapter 6**). Climates that were deteriorating or in some other way performing unnaturally gave cause for concern, since economic productivity, social stability and moral health could not be guaranteed. One dimension of 'good' colonial government was therefore the governance of climate. For example, Major George Chesney, a promoter of more direct British involvement in the government of India, remarked in 1877 that 'a study of the rainfall is one of the first duties of a civilised government' (cited in Anderson, 2005: 275). The climates governed by the European colonialists therefore became ones that were both 'weathered' *and* cultured; climates that encompassed both physical *and* imaginative properties.

As with the discourse of mid-latitude deforestation, unhealthy swamps and local temperature regulation, concerns grew among eighteenth-century island settlers about the impact of new plantation economies on (tropical) island rainfall. The colonial elites, including new cadres of state scientists, worried about the effects on local rainfall of forest and land clearing undertaken for new commercial plantations. This was especially true for insular colonies such as Mauritius, Saint Helena, Grenada and Barbados; these anxieties were expressed equally by British, Dutch and French governors. For example, in 1766 Pierre Poivre was appointed *commissaire-intendant* of Mauritius with a brief that included restoring the island's rainfall through forest conservation. The British East India Company voiced similar worries for the island of Saint Helena (Locher and Fressoz, 2012).

The colonial mission to civilise people and improve lands steered projects of climate governance in other directions. With insalubrious climates often to blame, attending to poor indigenous health and hygiene were frequently cited as justification for such projects. As with temperate climate governance, draining marshes and reforesting degraded land were proposed as ways of governing tropical climates and people to the betterment of both. As one French hygienist put it, 'colonising is sanitising' (Perier, 1845). And in a scheme pre-shadowing Sörgel's 1930s Atlantropa (see **Chapter 10**), the French colonialist François Élie Roudaire sought to make Algeria's desertic climate more temperate through inducing reliable rains. As a military officer in Egypt, Roudaire had noticed a humidifying of the climate around Suez after the Canal was completed in 1869 and he wanted to achieve a similar outcome for the French colony of Algeria. To steer Algeria's climate in this direction, in 1874 he proposed flooding a series of dry salt lakes (*chotts*) in the Sahara by digging a canal to the Mediterranean. Although never realised, governing climate in this way was to have improved Algerian agriculture; it formed part of the larger colonial 'regeneration mission' (Roudaire, 1874).

Colonial climate governance was not simply motivated by high ideals of beneficent paternalism, it was also intended as a demonstration of imperial power and of economic mastery. 'Conquering' colonised climates was often a proxy for the subjugation of colonised peoples (Harrison, 1996), and the economic motivation for governing 'unaccommodating' tropical climates was never far behind. For example, in the middle decades of the twentieth century British colonialists in east Africa found themselves in protracted struggles with a regional climate that could never be relied on. Relied on, that is, to produce the volume or reliability of agricultural yields demanded by investors or the institutions of imperial government. The British Overseas Food Corporation therefore invested heavily in surveys of these unruly climates – an essential step in objectification – but also made moves to govern local rainfall through piloting cloud-seeding technologies (Mahony, 2016). The ignominious failure of the 1940s Tanganyikan groundnut scheme demonstrated that merely the desire and ambition to govern climate is no guarantee that the atmosphere will cooperate.

Governing colonial climates – whether in practice or in the imagination – was often integral to much larger imperial projects of civilising, commercialising, sanitising and bringing orderly administration to unruly peoples and places. Even if the climates of these territories were 'steered' to some degree, this was only accomplished by controlling other aspects of the physical environment or other dimensions of social, economic or cultural life. Governing climate always expresses a political ambition which emerges from dominant power relationships, which is expressed through particular ideologies and which manifests in effects on specific people and places.

## Governing Global Climates

The same two principles of governmentality that I introduced earlier – objectification and problematisation – can be seen at work over the last 30 years as the project of *global* climate governance has unfolded. By objectifying global climate system behaviour through enumeration and mathematical simulation (see **Chapter 3**), science has placed (global) climate in the line of various other projects of high modernism. As James Scott explains in his book *Seeing Like a State*, governments of the modern era have increasingly sought to exercise their powers of regulation and control through enumeration of an expanding array of subjects. As Scott explains, formal state-sponsored cartographies were (and still are) used to manage territorial disputes, national censuses were used (and still are) to discipline citizens and Gross Domestic Product (GDP) was used (and still is) to regulate national economies (Scott, 1998). The relationship between the state and climate has followed a similar pattern. A scientific and systematic account of global climate, as distinct from accounts of local weather, opened the way for climate to be governed by the state; or, rather, by the free association of states meeting as the UN. If climate is constructed to be 'global', it must be governed 'globally'. Knowledge and power therefore become tightly coupled. Political rationalities, which seem both natural and inevitable and which justify some ends rather than others, always emerge from particular forms of knowledge (see **Box 11.2**).

---

## Box 11.2: Objects of Modern Climate Governance

What exactly is this global climate entity to be governed and what are its conditions that are deemed desirable or undesirable? Over the last 30 years global-average surface air temperature has emerged as the dominant index for objectifying global climate. Originally constructed by climate scientists and modellers to aid them in their scientific investigations, global temperature has turned out to be a powerful and convenient object of climate governance. Just as a well-functioning economy became defined in the 1950s in terms of GDP, so the performance of climate has become defined in terms of global temperature. Just as economies are now to be governed to secure specified annual GDP growth rates, so global climate is now governed to limit the rise in global temperature to 'well below 2°C'. This subjugation of good climate–society relations to the behaviour of a singular global index carries repercussions, just as reducing economic policy to maximising GDP does. Certain governing actions become imaginable and executable; others less so. Human security and well-being become narrowly defined in terms of an indexed global climate and effective political leadership in the world is reduced to securing the 2°C target. Indexing the object of climate governance to global temperature also opens the way for certain forms of technological experimentation to be imagined, for example, the creation of a 'global thermostat' through climate engineering (Hulme, 2014a).

 And yet global temperature is not in fact a tractable object of policy. It is too abstract and underdetermined. More immediate precursor indexes of climatic performance are needed, and so the idea of a global carbon budget has, in recent years, become equally central to the governance of global climate. In this calculus emissions of carbon dioxide are traceable back to specific human activities in particular places and so an annual global carbon emissions target (e.g. 'no more than 40 billion tonnes of carbon') can operate as an effective object of governance. Governing global climate thus becomes a task in governing those activities which contribute carbon dioxide to the atmosphere. At a generic level this would include, for example, the production and consumption of fossil-based energy, forest clearance, soil degradation, marine and air transport. But with global climate objectified and problematised in *this* way, the targets of governance can be disaggregated into ever finer spatial scales and activity units. Thus regulating, *inter alia*, vehicle emissions, farm-tilling regimes, human fertility, cattle feedstocks, showering times, human diet, air-miles travelled, night-time lighting, landfill volumes, building standards and much more, all become enrolled in the greater project of governing global climate.

This extension of climate governance from local and colonial climates to the grander ambition of *global* climate governance can be seen as a nascent impulse in one of President J.F. Kennedy's last speeches before his assassination in November 1963. In an address to the Anniversary Convocation of the US National Academy of Sciences on 22 October 1963, Kennedy drew attention to the unknown effects on the global biological and physical environments of various new human technologies. Alongside chemicals and nuclear weapons he also mentioned weather modification technologies which could directly influence global climate. Recognising a growing tide of public concern about the unknown and potential damaging effects of technology, Kennedy called for 'responsibility' in government: 'The Government has the clear responsibility to weigh the importance of large-scale experiments to the advance[ment] of knowledge or to national security against the possibility of adverse and destructive effects', (Kennedy, 1963). For Kennedy, governing global climate was to be about scientists assisting governments to arrive at 'rational judgements' about deploying, monitoring and regulating technologies and then 'interpreting these issues to the public'.

Yet this approach to governing global climate – governing climate from the 'centre' or the 'top' so-to-speak – has been shown to be unachievable. (This too was an argument of James Scott's *Seeing Like a State* (Scott, 1998). Social, natural and cultural contingencies always overwhelm the ordering eye of power; projects of centralised vision and control are always doomed to become fantasies that fail). The original objective of the UNFCCC, signed in 1992, was to 'stabilise greenhouse gas concentrations in the atmosphere at a level that would prevent dangerous anthropogenic interference with the climate system'. For nearly 20 years, the treaties and instruments designed to deliver this objective sought to regulate state actions through a targets-and-timetables approach under a legally binding international regime. This project gave birth to the Kyoto Protocol in 1997, but ended in acrimonious failure at the Copenhagen negotiations in 2009.

What subsequently emerged in the 2015 Paris Agreement was something very different. Although reductionist temperature and carbon emissions targets were inscribed in the Agreement (2°C, 1.5°C, net zero global carbon emissions by the end of the century), everything else about the Agreement suggested that climate governance will in fact unfold from the 'periphery' or the 'bottom', to pursue the metaphor. National governments are asked only to bring their (voluntary) Intended Nationally Determined Contributions (INDCs) to the Agreement. There is no enforcement regime, merely pledge-and-review. The Agreement recognises the many loci of governance which exist beyond the nation state: businesses, municipalities, the courts, civic movements, finance and emissions markets.

On the one hand, the Paris Agreement could be interpreted as a demonstration of the dominant neo-liberal ideology of the early twenty-first century – the currency of numbers, market competition, deregulation, volunteerism, individualism. On the other hand, it concedes a very different conception of power and governance than did the Kyoto Protocol. Governing climate, it would seem to suggest, can never be a secure accomplishment (Bulkeley, 2016). Rather than being programmatic – 'accompanied by an eternal optimism that a particular domain or society could be administered better or more effectively, that reality is, in some way or other, programmable' (Miller and Rose, 1990: 4) – the Paris Agreement tacitly concedes that the outcome of governing global climate cannot be secured nor safely engineered. Neither state or UN-centric manifestations of power, nor the machinations of the market, nor the surprises wrought by cultural and technological shifts, will ever allow future climate a predictable 'safe-landing'. In other words, governing global climate is not about delivering a pre-determined outcome, but about incentivising different actors and interests to intervene consciously in the world with climatic concerns 'in view'. This is an optimistic reading of the Paris Agreement, one which emphasises the possibilities of an agonistic democracy. Climate in a formal sense is not ungovern*able*; there is *ability* to steer or to influence climate. But climate *is* uncontrollable, as in the social or engineering sense of securing a particular end-state.

Expressing this differently, it is possible to conceptualise global climate governance as a project that is always doomed to fail. The initial construction in the 1980s of global climate as an object of governance was accompanied by a narrative of it as already being broken or damaged, of global climate already being changed for the worse as a result of human activities. There has never been human knowledge of a healthy or 'normal' conceptually *global* climate, as opposed to knowledge of the varieties of local climates within which cultures have traditionally evolved in the past. In contrast, the concept of global climate is one that 'is at its conceptual root *always already broken*, thereby engendering failure in a Sisyphean quest to fix what is conceptually unfixable' (Hamilton, 2015: 141, emphasis in original). According to this reasoning, governance of global climate from the centre or the top can never succeed; a solution for climate-change can never be found.

Yet this does not mean that the idea of global climate has no purchase in the political world. If climate is understood more as a metaphor for the ordering of human life, its power being to reflect and to shape the ways in which humans live together – how they live culturally with their weather – then climate will continue as an object of contestation among different social actors. After all, there are many competing visions of how human life is best ordered. In which case, the idea of governing *global* climate, as a surrogate for the idea of ordering *global* society, makes climate even more expressly an object of politics. As I have shown in earlier chapters, the idea of climate connects all things together; for example, science, markets, identity, wealth, health and power. It has done so in the past and it continues to do so today. Ambitions to govern global climate cannot but convert diverse socio-technical objects and cultural practices and institutions into sites or modes of political argument. Thus, *inter alia*, food, cars, babies, music events, airports, cookers, waste, art exhibitions, wind turbines and educational curricula become contested with regard to whether or not they are 'climate-friendly', enlarging the repertoire of argumentation about what are already political objects.

Rather than the idea of climate-change *being* politicised (by nefarious actors), global climate is an idea which is political by construction. With global climate as an object of governance, climate-change cannot but be an idea which mobilises powerful actors in pursuit of their different interests. Under this condition, and with the danger lurking that a totalising object of governance can justify a totalisation of unaccountable power, one might argue that the means matter more than the end. Re-securing global climate should not be delivered at any cost, not least at the cost of hollowing-out or dismantling the democratic virtues of pluralism, agonism and accountability.

## Chapter Summary

In order to govern it is necessary to specify the object of governance (e.g. a society or a business) and to have clear and specific goals for this object (e.g. social stability or a profitable business). In order to govern well it is also necessary for those governing institutions and agents to have access to means to achieve these goals (e.g. taxation or employee incentives) and to possess appropriate knowledge of how the object of governance responds to instruments and policies (e.g. political economy or double-entry accounting). These principles of governmentality, originally identified by the French philosopher Michel Foucault in the 1970s, can also be applied to the governing of climates.

In this chapter I have given some examples of how these principles have been directed towards the governing of different climates, whether local, colonial or global. Yet none of these dimensions of governmentality is straightforwardly applicable to the idea of climate. As earlier chapters in the book have shown, reducing climate to a mere physical entity does not do justice to the imaginative power of climate as an idea. So when climate is to be governed what precisely is it that is being steered? The cyclones, heatwaves, ice-storms and downpours that begin to constitute the physical and imaginative contours of climate are not directly subject to human laws, policies or technologies. Governing climate therefore always becomes a project about governing and controlling things other than the weather: physical environments, social practices, material technologies, investment flows. This can be seen in the examples offered above, where the draining of marshes, the planting of trees or the redesigning of cities have been activities embarked on at different times and in different places as projects in governing local climates. Governing local or colonial climates has usually been an exercise in governing local communities or colonial societies.

The rise of global climate governance in recent decades has further extended the range of practices, technologies and institutions which can come under the reach of climate governance. The object of governance as specified in the Paris Agreement is global temperature and while this has indeed been problematised – a rise of 2°C or more above pre-industrial temperature is deemed by the world's governments to be dangerous – global temperature is not an object which is directly tractable to human actions. Governing global temperature therefore requires, at the least, governing the full range of human activities and technologies – and the imaginations which give rise to them – which emit greenhouse gases and other particulates into the atmosphere. This in turn requires virtually every human practice becoming subject, at least in principle, to the logic of global climate governance. Land, energy, mobility, diet, forests, procreation and, in the end, all human behaviour become subject to the totalising idea of climate governance.

Governing global climate becomes an exercise in governing global society, but where the power to do so exists in no central or identifiable locus. Society, and thus climate, is uncontrollable in the sense of engineering a pre-determined social or physical outcome. And yet in a different sense, society, and therefore climate too, is being eternally governed, but in many different places, through different social actors and with indeterminate outcomes.

In the final chapter of the book I turn more explicitly to the future of climate: its future as an idea, more than its future as a material system.

## Further Reading

Bäckstrand, K. and Lövbrand, E. (eds) (2015) *Research Handbook on Climate Governance*. Cheltenham: Edward Elgar.

Bulkeley, H. (2016) *Accomplishing Climate Governance*. Cambridge: Cambridge University Press.

Bulkeley, H. and Newell, P. (2010) *Governing Climate Change*. Abingdon: Routledge.

Gupta, J. (2014) *The History of Global Climate Governance*. Cambridge: Cambridge University Press.

# 12

# Reading Future Climates

## Introduction

Historians are not known for telling stories about the future. They are usually more interested in constructing tales about the past. Historical 'facts' are woven together to provide convincing accounts of 'how one thing led to another', seeking insights into why people acted the way they did and with what consequence. But there are no facts about the future for historians to discover or construct. The future is usually left to the imagination of novelists or inventors, to the visions of seers or to the predictive models of scientists.

In a mini-book published in 2013, *The Collapse of Western Civilisation: A View from the Future* (Oreskes and Conway, 2014), historians Naomi Oreskes and Erik Conway did turn their attention to the future, the climatic future. They adopted the literary trope of science fiction to imagine the Earth's future climate from the perspective of the year 2373 CE. Adopting the mantle of future Chinese historians, Oreskes and Conway looked back on what they described as a 'second Dark Age', prompted by the Great Collapse and Mass Migration of 2074. In telling this story they recounted the climate events of the twenty-first century as they appeared to these imagined historians of the Second People's Republic of China. What emerges is an imaginative tale about the future, written from the future: a 'historical' account of climate and humanity during the twenty-first century. It is a tale of accelerating climatic chaos, social disruption and human catastrophe. The mass fatalities of the domestic cats and dogs of wealthy westerners in 2023 is only the start.

By 2042 the global-average surface air temperature has risen nearly 4°C and solar climate engineering commences in 2052 ... and stops again in 2063. This sudden cessation of sulphate aerosol injection causes the greenhouse effect to reach a 'global tipping point' by the mid-2060s. Massive releases of Arctic methane push global temperature 11°C higher than today. What follows combines Cormac McCarthy's post-apocalyptic novel *The Road* with the myth of the Noahian Flood. Sea-level rises 7 metres between 2074 and 2093 (a rate of 1 ft. per year), 1.5 billion people are made climate refugees and 60–70 per cent of species are extinguished. The entire human populations of Australia and Africa (over 4 billion according to the latest UN projections) are wiped out.

Like science fiction, futurist history is an imaginative literary trope which can be used to entertain, to speculate or to offer moral instruction. In *The Collapse of Western Civilization* Oreskes and Conway engaged in this latter pursuit. They were explicit about the root historical causes of their civilisational collapse: the 'two inhibiting ideologies: positivism and market fundamentalism' (2014: 35). In weaving together science, moralism and the human imagination they offered an excellent example of how human thinking about the climatic future is always culturally shaped. Oreskes and Conway's deep antipathy to positivist science and to market neo-liberalism shaped their declensionist account of western civilisation and its devastating effect on global climate. Science in the twenty-first century was, they claimed, too reductionist, too gendered and too distorted by disciplinary practices. Physical scientists practised 'intellectual self-denial', in thrall to constricting conventions of false statistical rigour, or else they embarked on 'arcane arguments' about extreme weather-event attribution. According to Oreskes and Conway, science appropriated too much public resource to the detriment of public investment in the arts. Their sagacious Chinese historians were startled that 'these western people' had not translated knowledge into power; they lamented that power had been in the hands of political institutions and interests rather than in the hands of scientists. It was only heroic individual scientists who had sought to overcome nefarious and obstructive western politicians, corporations and citizens.

Cultural theorist Arjun Appadurai has argued that humans can only know the future culturally, that the future is a 'cultural fact' (Appadurai, 2013). Consequently, and inevitably, there are many possible futures; and it follows too that there are many possible climate futures. Scientists and prophets – or indeed historians such as Oreskes and Conway – might offer their predictions of future climate, but, given that the future always recedes ahead of us, today's predictions will always be supplanted by those of tomorrow. Today's climate is always yesterday's future climate as predicted by past prophets. For Appadurai, how people imagine the future reveals a lot about their political beliefs and their cultural practices of the present and past. And this is as true for the climatic future as it is for other futures, whether economic, technological, social, medical. Imagined or predicted climates of the future might evoke emotions of fear, loss and nostalgia through dystopic accounts, but they might also have the potential to engender faith in human capabilities to bring about a better or a more just future. As I showed in the previous chapter, attempts to govern climate become battlegrounds upon which different political visions of the 'well-ordered society' or the 'good life' vie for supremacy.

*\* \* \**

In this concluding chapter I engage with some of the ways in which future climates are being imagined at the present time. I suggest three patterns identifiable in these imaginative worlds. The desire to re-secure future climate is one and the recognition that future climate will always be in a state of improvisation is another. But there is also a third possibility: that in the epoch of the Anthropocene (see **Box 12.1**) the very idea of climate as a useful ordering device for humanity is moot.

## Box 12.1: Climate in the Anthropocene

At one level, the idea of the Anthropocene is disarmingly simple. It simply captures the idea that humans, although by no means all humans equally, leave their mark on the physical world at a planetary scale. Yet it is also an idea pregnant with multiple meanings, many of them yet to be imagined. 'The Anthropocene' is a term popularised during this century's first decade by Earth system scientists Paul Crutzen and Eugene Stoermer (2000) to describe a new geological epoch in which humanity now lives. They argued that the current geological epoch – the 12,000-year-old Holocene – has now been exited and a new epoch should be recognised, one in which the human species possesses Earth-shaping force.

The idea of the Anthropocene is fruitful and creative, by no means limited merely to its labelling of the latest planetary epoch. The Anthropocene provokes a re-imagining of the place and purpose of human life and action and how people see themselves in relation to the non-human world. Rather than being limited to a label for an epoch, the Anthropocene is more usefully understood as an ideology. It provides the ideational underpinning for a particular view of the world in which humans' creative powers are central (see **Figure 12.1**). It is closely related to the ideas of the post-natural (nature as no longer natural) and the post-human (humans as more-than-human). The Anthropocene imaginary also questions the suitability of conventional science, and its categorisation of the world, to generate adequate knowledge of this new state of affairs. The methodological separation of object from subject would appear to deny one of the characteristic attributes of an Anthropocenic world, namely that humans and non-humans are conjoined. In the Anthropocene the scientist is no longer studying the workings of nature, but the co-products of a nature already studied and acted upon (and acted upon and studied) recursively by humans. There is no independent vantage point from which to pronounce 'discoveries of nature' for they are discoveries of something else – of socio-natures or of 'more-than-natures' which already bear the mark of their discoverer.

But the era of the Anthropocene also challenges humanities and social sciences disciplines. There is no natural, unadulterated stage upon which humanity treads, upon which human history is played. Ethics, politics, theology, history, literature need to adjust to the new materialisms of 'vital matter' (J. Bennett, 2011) which are now in play.

**Figure 12.1**   The idea of the Anthropocene has transcended academic disciplines, being widely discussed and debated in sciences and humanities alike. This poster advertises the Interdisciplinary Humanities Center's 2014–2015 public events series at the University of California at Santa Barbara: 'The Anthropocene: Views from the Humanities' (Source: http://www.ihc.ucsb.edu/series/anthropocene/ (accessed 19 May 2016)).

## Re-Secured Climate

As I have shown in the previous chapters, people's engagement with the idea of climate has always had physical and psychological dimensions. A disturbance to the stability of physical climate, whether real or perceived, has a disconcerting effect on the human sense of security and well-being, both materially and emotionally. Any disaster changes the lifeworld of an individual, but when a series of climatic disasters are strung together so that they tell of 'a change in the climate' the feelings of loss and estrangement are all the more powerful. Explanations for such unsettling of the natural orders people construct for themselves are frequently sought in the realm of human behaviour. As I showed in **Chapter 4**, climate instability has – from the myth of Noah's Flood onwards – long functioned both as a sign of, and a warning against, human misdemeanours. The realisation of a change in climate has often been seen as a moment for moral cleansing or personal transformation, captured by the Greek word *kairos* – meaning the right or opportune moment[1]. One of the

---

[1] In modern Greek *kairos* also means 'weather', which adds another layer of meaning to this suggestion.

lessons Oreskes and Conway (2014) sought to convey in their morality tale about the climatic future was that in order to prevent a radical destabilisation of the world's climate, people of today's generation should change their ways. Although writing as imagined Chinese historians of the far future, the targets of their reproof were you and me, here and now, particularly those of us giving succour to those twin ideologies of 'positivism and market fundamentalism'.

The American fantasy drama television series *Game of Thrones* (GOT) illustrates another way in which an impending change in climate works metaphorically as existential threat and as a call to repentance and change. The television series is based on the series of epic fantasy novels *A Song of Ice and Fire* by George R.R. Martin in which the fictional peoples of the Land of Westeros and the city of King's Landing seem threatened by a deteriorating climate. The zombie-like White Walkers are harbingers of a devastating winter climate, confronted only by the Night's Watch, black-clad warriors who seek to convince the residents of King's Landing of this imminent climatic threat. In GOT climatic change functions as a metaphorical but perpetual danger, the risks of which – disrupted agriculture and food supplies, personal discomfort – threaten the stability and survival of the political order. In commenting on the active bloggers who engage with the TV series, Manjana Milkoreit argues for the crossover between winter in Westeros and how the threat of a deteriorating climate future functions in the real world (quoted in Judd, 2015). Winter is coming, but few in Westeros seems to believe it; the world is warming, but American politicians are divided or apathetic. One can also see in GOT a parallel to Oreskes and Conway's *The Collapse of Western Civilisation*: the devoted Night's Watch stand for Oreskes and Conway's heroic individual scientists, seeking to warn a sleeping world about an oncoming threat.

The aim of these morality tales, whether Oreskes and Conway's fictional history or *Game of Thrones*, or indeed today's goals in tackling climate change, is to re-secure climate through some combination of governmental, technological, social or personal transformation. For example, through the collective effort of nations, the Paris Agreement on Climate Change (see **Chapter 11**) seeks to stabilise climate at no more than between 1.5 and 2°C warmer than the pre-industrial level; or as summarised by the journal *Nature*, '... a global deal to *secure* a safe ecological future for all' (Anon, 2015; emphasis added). The discourse of climate engineering (see **Chapter 10**) points rather to novel technologies which might re-balance planetary flows of energy and carbon to deliver a stable and secure climate. Through a radical restructuring of social and economic power structures, Naomi Klein (2014) seeks to dismantle the carbon capitalism that has destabilised global climate, while scholars such as Karen O'Brien draw attention to the potential of personal transformation to contribute to a re-securing of climatic and cultural life (O'Brien, 2013).

These instincts to re-secure climate within desirable and 'safe' limits, to bring climate back into some sort of orderly condition, are all recognisably human. If a stable climate is deemed to be a global public good, if it has only been the benign climate of the Holocene which has allowed human civilisation to flourish, if human intrusion in the atmosphere is regarded as despoiling a natural and desirable state ... if all these things are true, then it may well seem logical and necessary for global climate to be rehabilitated, to prevent it from becoming too disorderly. Such impulses to re-secure and protect climate speak to the enduring human need for order and stability (Daston, 2010). Even if it were possible to live materially with the effects of a changing climate, it might not be psychologically desirable to do so.

## Improvised Climates

Much thinking about the future of climate is therefore driven by the desire to guard and defend climate against the perceived distorting effects of human actions; to eradicate – or at least to minimise – the 'unnatural' influences of humanity on climate. The instincts described above are bound up with the value people perceive of retaining climate as an idea which restrains the human experience of weather from descending into chaos. This indeed is a powerful function of the idea of climate which, throughout this book, I have argued can be found across different historical and geographical cultures.

But a rather different imaginative stance with respect to the climatic future embraces the idea of improvisation, the idea of creating something unplanned and ad hoc out of materials immediately to hand. In this line of thought, a changing climate *also* functions symbolically, but rather than as a signifier of cultural anxiety or moral culpability, a changing climate becomes a metaphor for the spirited life and irrepressible creativity of humanity. Thinking of future climate as something that can, and will, only result from human improvisation requires starting from a different premise to that which provides the impulse to re-secure climate, to bring climate back into order. Yes, the premise for improvisation recognises the deep material traces left on the physical planet by human cultural practices. In this sense, the epoch of the Anthropocene (see **Box 12.1**) is indeed suitably named given the significant influence exerted on the physical world by multiple human actions. But the starting premise for improvisation is one which diverges from the premise inspiring the climatic projects of transformation mentioned in the previous section on this one important point. Rather than treating climate as being controllable through political will, technological ingenuity or moral integrity, or even through divine oversight, climate 'improvisers' recognise that the physical processes of the planet's atmosphere and oceans will always escape bounded human efforts to constrain them (Clark, 2011).

This contrast between human agency as having material force on the physical world and physical processes which stubbornly escape human control sets up the tension which characterises how my improvisers might think of the climatic future. Future climates will not be securable in the old modernist sense of mastery, design or purposeful governance. There can be no going back to the climatic past, nor any possibility of maintaining a secure, stable or even predictable future climate through directed interventions. A natural atmosphere, whether real or imagined, can be no more; oceans are being carbonated, ice-sheets set in motion, glaciers in retreat. And yet people, collectively, are neither passive nor powerless. Given Anthropocenic possibilities, my improvisers would argue that people – as actors with global reach – cannot *but* take on a conscious and reflexive role in relation to the climatic future. Having dispensed with God and having re-joined nature, only the option of taking on an active yet humble responsibility for the coming climate remains open to people of the twenty-first century. But the outcomes of such responsibility will be severely circumscribed and unpredictable. Neither scientific knowledge nor a semi-domesticated nature will allow humans to guarantee the climatic outcomes of their behaviours and technologies. Whether one thinks about human technological actions performed on future climate as enhancement or as restoration (**Chapter 10**), these metaphors are equally suggestive of powers that are, in fact, denied to humans.

In the terrestrial sphere the recent vogue for re-wilding might be seen as one expression of this ethic of improvisation: rearranging nature from whatever spaces and species are readily available, but without a prepared masterplan or secure knowledge of the outcome. Re-wilding might enthusiastically be embraced as an attempt to recreate pre-modern ecosystems, but such an ambition can never be achieved. Contexts are different, nature does not cooperate, unforeseen side-effects occur. Improvised adjustments to local ecologies are made continuously by both non-human and human actors, the results of which are novel ecosystems and new species assemblages.

So too with the atmosphere. The ambition to return the atmosphere to pre-industrial conditions, or even to re-secure a stable climate no warmer than 2°C above pre-industrial levels, may in some ways be conceived as 're-wilding' the climate. But such efforts are doomed to fail. Contexts are different, nature does not cooperate, unforeseen side-effects occur. In the Anthropocene the only possibility is the creation through improvisation of new and wholly underdetermined climatic futures. What will be created will be novel climates with new assemblages of local weather. Rather than re-creating a climatic past, the only possibility is to go forward with new sets of conditions in place. The interplay between the material emanations of humanity's cultural evolution and the physical forces of the non-human world will lead to perpetually improvised climates, climates which are neither stable nor predictable. Or to use the language of Bruno Latour, future climate will be 'composed' through an indescribable array of multiple human actions. It will be recognised that 'this common [climate] has to be built from utterly heterogeneous parts that will never make a whole, but at best a fragile, revisable, and diverse composite material' (Latour, 2010: 474).

In similar vein, sociologist Bron Szerszynski resists the argument than future climate should or even can be brought under (human) technical control. There will always remain a powerful 'otherness' to the weather, he says. It will never be tamed by humans' cultivating powers, just as in the past it never was tamed by supplications to the gods or through the protective idea of a stable climate (**Chapter 4**). The weather will always escape such human efforts. Instead, a more humble acceptance is needed of the new improvised and experimental relationship with climate that is now created (Szerszynski, 2010). This requires a different understanding of the human relationship with nature (see **Box 12.2**).

---

## Box 12.2: Weatherculturalists

Several writers have used the practice of gardening as a means for re-thinking human attitudes towards nature in the Anthropocene. Applying this metaphor to the future of climate moves the emphasis away from ambitions to re-secure climate through modernist projects of control, to thinking more in terms of improvisation, of working with nature to fashion outcomes which are neither fully predictable nor fixed. As geographer Holly Buck explains, 'The garden is a site through which we can examine connection and care in practice. It is a powerfully enchanting trope: the linguistic enchantment of the garden of love, the walled garden, the secret garden, and so on. The Anthropocene provokes the question of scale ... large-scale industrial monocropped landscapes are a referent for Anthropocene horror tales; planetary gardening imagines something quite different' (Buck, 2015: 374).

Gardeners require the virtues of humility, cheerfulness, attentiveness and mindfulness as they go about their work (di Paola, 2015). Gardens are, of course, a joint product of human imagination and skill working with and through processes of soil conditioning, photosynthesis and the weather. In her relationship with nature, a gardener is neither in control nor powerless. There is a mutuality in which – at least in the best gardens – human vision and intention can find expression, alongside a celebration of the freedoms possessed by plants, animal life and soil. The battle against weeds is relentless and without end, but is an activity freely engaged in by the gardener and pursued resolutely. The focus is on the practice of gardening 'rather than on its outcomes and on the benefits internal to the practice rather than those produced by it' (2015: 203). Gardening becomes a metaphor for caring and for making mindfully and responsibly, virtues that are needed for composition in the Anthropocene. But human cultivating practices have yielded not just gardens. Over countless generations they have also yielded agricultures, horticultures, aquacultures, silvicultures and permacultures. Is it fruitful, is it possible even, to think in terms of weathercultures and people as weatherculturalists?

---

## Post-Climate

There is a third imaginative stance which may be taken in relation to the future of climate, one which pushes the idea of improvisation still further. The argument here runs as follows. Human cultures and atmospheric weather are becoming increasingly inseparable. Each is shaping the other in that what humans do shapes the skies overhead, while cultures are weathered through climate. In the Anthropocene, people have moved from symbolically creating their many small worlds to materially co-creating their one entire world. Or rather, their many small symbolic worlds are now embedded within a much larger co-created material world which itself carries cultural symbolism. Historian Dipesh Chakrabarty has argued that one result of Anthropocenic thought is that the historian's distinction between natural and human histories has begun to collapse (Chakrabarty, 2009). Similarly I argue that the distinction between human and natural climates is no longer meaningful.

The central idea of the Anthropocene is that *change* is now inescapable and perpetual. There is no normal. But this is not merely because natural processes are changing material forms; such has been the case throughout the course of Earth's deep history. This is because material forms are now increasingly bound up on a global scale with human processes of discovery, invention and improvisation. In other words, changes in the material world now emerge from the irrepressible human technologies and practices which originate in the cultural imagination. Whether it is human bodies, material technologies, urban ecologies or regional climates, nothing is now merely natural or gifted to the human, other than existence itself. *Climatic* change – i.e. change that is defined by the adjective 'climatic' – is therefore losing any distinct meaning. Changes occurring to physical climates can no longer be isolated from changes occurring to human economies, technologies, societies and cultures. As offered by Margaret Atwood (see **Chapter 1**), 'I would rather call [climate-change] the *everything change*' (see also **Box 1.1**). 'Climatic change' is simply a meta-category of change, an aggregated manifestation of changes which are at one and the same time environmental, economic, technological, social and cultural.

In this train of thought, the 'normal' historical function of the idea of climate – to stabilise relationships between weather and culture – is moot. The Anthropocene suggests the possibility of such stability is a chimera. Climate can no longer be helpful as an idea that sits between weather and culture because weather and culture are fusing into a single reality with no independent mediator; we are the weather and the weather is us, an idea played on by Roni Horn in her art project *Weather Reports You* (Horn, 2007). Climate is therefore becoming everything, but also nothing. It makes no sense to speak of climate, when the imaginative work performed by the idea of climate no longer has cultural traction. Rather than being useful as an imaginative way of, first, separating weather and culture and, then, of stabilising relationships between them, climate may become a zombie concept (Beck and Beck-Gernsheim, 2001) – an idea which is apparently dead, but which continues to 'live-on' through its intellectual and imaginative legacy. Metaphorically speaking, the climate of the Anthropocene is becoming climate-less.

# Summary

In *Weathered: Cultures of Climate* I have shown how the idea of climate has worked historically to stabilise cultural relationships between people and their weather and how it continues to do so; at least until now. The idea of climate has continually readjusted to changes both in atmosphere and weather as well as in human culture. Climate has been productive in that it has enabled people to live with their fickle weather through a changing and widening range of cultural resources, practices, artefacts, myths and rituals. As part of this story I have shown how the recent phenomenon and discourse of climate-change is grasped and represented culturally. Climate-change should not be understood as a decisive break from the past nor as a unique outcome of modernity; it should be seen as the latest stage in the cultural evolution of the idea of climate. How societies respond to climate-change is simply the latest manifestation of the human search for ordering nature–society relations.

But now, in this last chapter of the book, I have speculated about the future of climate as an idea. I have offered three possibilities. First, and most conventionally, is the idea of climate re-secured within desirable and 'safe' limits, of eliminating or at least minimising the effects of human influences on the physical climate of the future. This ambition seeks to shore up the historical function of climate by re-establishing a degree of orderliness in the world. A second possibility recognises the limitations of this impulse and embraces a future of improvised climates rather than of re-secured or stabilised ones. Improvisation accepts a more humble disposition with regard to the relative powers of the human and non-human world. It nevertheless recognises the inevitability of purposeful human actions re-making the climate, but only within certain limits of possibility. Physical climates will always escape human management. This stance requires some re-evaluation of the imaginative force of climate as a stabilising idea.

The third possibility is more radical and calls into question the imaginative function of climate upon which this book is premised. In the new era to which the idea of the Anthropocene seeks to give expression, perhaps people will have to learn to live without the idea of climate. At least learn to live without climate as an idea that brings order and stability to relationships between weather and culture. Through cultural imaginaries driven by metaphors such as 'tipping points', narratives such as the Anthropocene and technologies such as Google Earth, the climatic future is already being composed. Before predicted climates come to pass, the imaginative contours of their form will already have changed. And will do so again. And again. Given my argument that all human cultures are weathered and that scientific investigations show that weather is increasingly being cultivated, new and more fruitful categories than simply 'climate' are needed to work with. In the Anthropocene there can be no climate in the old sense; only weathercultures, with people acting as weatherculturalists.

Far from being able to articulate, define and set course for the 'climate we want' (Caseldine, 2015) – whether 1, 2 or 3°C of warming – the only task left is to think metaphorically about climate. In the Anthropocene the question to be answered is what is the *metaphorical* climate 'we want'? What are the political, intellectual, economic, cultural and moral climates to be cultivated? Because, in the end, it is *these* metaphorical climates that will arrange the weathercultures of the future. These weathercultures will not be as objects of governance, but as outcomes of actions taken by certain types of people and not others.

Maybe then, the idea of climate has already served its purpose. Maybe the human condition has outgrown the usefulness of climate as an idea which stabilises and sustains human life. The 'new normal' of climate is simply that there can be no normal. And this is unsettling. New concepts and metaphorical ideas, beyond climate, will be needed which meet the emotional, spiritual and material demands of living in an atmosphere and with the weather it yields which, without respite, are now both of human making.

## Further Reading

DeFries, R. (2014) *The Big Ratchet: How Humanity Thrives in the Face of Natural Crisis.* New York: Basic Books.

Flannery, T. (2015) *Atmosphere of Hope: Searching for Solutions to the Climate Crisis.* London: Penguin/Random House.

Jamieson, D. (2014) *Reason in a Dark Time: Why the Struggle Against Climate Change Failed – and What it Means for Our Future.* Oxford: Oxford University Press.

Williston, B. (2015) *The Anthropocene Project: Virtue in the Age of Climate Change.* Oxford: Oxford University Press.

# Bibliography

Adamson, G.C.D. (2012) '"The languor of the hot weather": Everyday perspectives of weather and climate in colonial Bombay, 1819–1828', *Journal of Historical Geography* 38 (2), 143–54.

Adamson, G.C.D. (2015) 'Private diaries as information sources in climate research', *WIREs Climate Change* 6 (6): 599–611.

Agarwal, A. (1995) 'Dismantling the divide between indigenous and scientific knowledge', *Development and Change* 26: 413–39.

Agnew, J. (2014) 'By words alone shall we know: Is the history of ideas enough to understand the world to which our concepts refer?', *Dialogues in Human Geography* 4 (3): 311–19.

Anderson, J. (2015) *Understanding Cultural Geography: Places and Traces,* (2nd edition). Abingdon: Routledge.

Anderson, K. (2005) *Predicting the Weather: Victorians and the Science of Meteorology.* Chicago, IL: Chicago University Press.

Anon (1891) 'Weather tinkering', *New York Times*, 23 September, p.4.

Anon (1958) 'Climate control is coming', *Newsweek* 13 January issue, http://blog.modernmechanix.com/climate-control-is-coming/ (accessed 16 May 2016).

Anon (2008) 'Gore's Götterdämmerung', *Science* 320 (5882): 1401. DOI: 10.1126/science.320.5882.1401c.

Anon (2015) 'Future proofing', *Nature* 528: 164.

Appadurai, A. (2013) *The Future as Cultural Fact: Essays on the Global Condition.* London: Verso.

Barnes, J. and Dove, M.R. (eds) (2015) *Climate Cultures: Anthropological Perspectives on Climate Change.* New Haven, CT: Yale University Press.

Barnett, L. (2015) 'The theology of climate change: Sin as agency in the Enlightenment's Anthropocene', *Environmental History* 20: 217–37.

Barry, A. (2001) *Political Machines: Governing a Technological Society.* London and New York: The Athlone Press.

Beattie, J. (2009) 'Climate change, forest conservation and science: A case study of New Zealand, 1860s–1920s', *History of Meteorology* 5: 1–18.

Beck, U. (1992) *Risk Society: Towards a New Modernity.* London: Sage.

Beck, U. (1997) 'Global risk politics', in M. Jacob (ed.), *Greening the Millennium? The New Politics of the Environment.* Oxford: Blackwell. pp. 18–33.

Beck, U. and Beck-Gernsheim, E. (2001) *Individualization: Institutionalised Individualism and its Social and Political Consequences.* London: Sage.

Behringer, W. (1999) 'Climatic change and witch-hunting: The impact of the Little Ice Age on mentalities', *Climatic Change* 43 (1), 335–51.

Behringer, W. (2010) *A Cultural History of Climate*. Cambridge: Polity Press.

Bennett, B.M. (2011) 'Naturalising Australian trees in South Africa: Climate, exotics and experimentation', *Journal of South African Studies* 37 (2): 265–80.

Bennett, J. (2011) *Vibrant Matter: A Political Ecology of Things*. Durham, NC: Duke University Press.

Blanchard, J. (2015) 'Climate Change: A Rewired Food Web', *Nature* 527: 173–174.

Boia, L. (2005) *The Weather in the Imagination*. London: Reaktion Books.

Bölsche, W. (1919) *Eiszeit und Klimawechsel* (Kosmos Gesellschaft der Naturfreunde). Stuttgart: Franckh'sche Verlagsbuchhandlung.

Bonacina, L.C.W. (1939) 'Landscape meteorology and its reflection in art and literature', *Quarterly Journal of the Royal Meteorological Society* 65 (282): 485–98.

Botero, C.A., Gardner, B., Kirby, K.R., Bulbulia, J., Gavin, M.C. and Gray, R.D. (2014) 'The ecology of religious beliefs', *Proceedings of the National Academy of Sciences* 111 (47): 16784–9.

Bottoms, S. (2012) 'Climate change "science" on the London stage', *WIREs Climate Change* 3 (4): 339–48.

Brace, C. and Geoghegan, H. (2011) 'Human geographies of climate change: Landscape, temporality, and lay knowledges', *Progress in Human Geography* 35 (3): 284–302.

Bristow, T. and Ford, T.H. (eds) (2016) *A Cultural History of Climate Change*. Abingdon: Routledge.

Broecker, W.S. (1987) 'Unpleasant surprises in the greenhouse', *Nature* 328: 123–6.

Brönnimann, S. (2002) 'Picturing climate change', *Climate Research* 22 (1): 87–95.

Brückner, E. (1890) 'Klimaschwankungen seit 1700 nebst Bemerkungen über die Klimaschwankungen in der Diluvialzeit', *Geographische Abhandlungen* 4: 153–484. Available from https://archive.org/details/bub_gb_UCRUAAAAMAAJ (accessed 6 May 2016).

Bryson, R.A. and Murray, T.J. (1977) *Climates of Hunger: Mankind and the World's Changing Weather*. Madison, WI: The University of Wisconsin Press.

Buck, H. (2015) 'On the possibilities of a charming Anthropocene', *Annals of the Association of American Geographers* 105 (2): 369–77.

Buell, F. (2004) *From Apocalypse to Way of Life: Environmental Crisis in the American Century*. Abingdon: Routledge.

Buffon, G.-L.L. de (1778) *Histoire naturelle, générale et particulière*, vol. 20: *Époques de la nature*. Paris: Imprimerie Royale.

Bulkeley, H. (2016) *Accomplishing Climate Governance*. Cambridge: Cambridge University Press.

Bulkeley, H. and Stripple, J. (2015) 'Governmentality', in K. Bäckstrand and E. Lövbrand (eds), *Research Handbook on Climate Governance*. Cheltenham: Edward Elgar Publishing. pp. 49–59.

Burke, E. (1968[1757]) *A Philosophical Enquiry into the Origin of Our Ideas of the Sublime and Beautiful*, J. Boulton (ed.). South Bend, TX: University of Notre Dame Press.

Burke, M.B., Miguel, E., Satyanath, S., Dykema, J.A. and Lobell, D.B. (2009) 'Warming increases the risk of civil war in Africa', *Proceedings of the National Academy of Sciences USA* 106 (49), 20: 670–4.

Burke, M., Hsiang, S.M. and Miguel, E. (2015) 'Global non-linear effect of temperature on economic production', *Nature* 527: 235–9.

Burke, P. (2008) *What is Cultural History* (2nd edition). Cambridge: Polity.

Calhoun, C. (2008) 'A world of emergencies: Fear, intervention and the limits of cosmopolitan order', *Canadian Review of Sociology* 41 (4): 373–95.

Cameron, D. (1964) 'Early discoverers. XXII: Goethe – discoverer of the Ice Age', *Journal of Glaciology* 5 (41): 751–4.

Carey, M. (2011) 'Inventing Caribbean climates: How science, medicine, and tourism changed tropical weather from deadly to healthy', *Osiris* 26: 129–141.

Carlsson, A. (2009) 'What is a storm: Severe weather and public life in Britain in January 1928', in V. Janković and C. Barboza (eds.), *Weather, Local Knowledge and Everyday Life: Issues in Integrated Climate Studies*. Rio de Janeiro: MAST. pp. 87–98.

Caseldine, C. (2015) 'So what sort of climate do we want? Thoughts on what is "natural" climate', *The Geographical Journal* 181 (4): 366–74.

Castree, N. (2013) *Making Sense of Nature*. Abingdon: Routledge.

Chakrabarty, D. (2009) 'The climate of history: Four theses', *Critical Inquiry* 35 (Winter): 197–222.

Clark, N. (2011) *Inhuman Nature: Sociable Life on a Dynamic Planet*. London: Sage.

Corton, C.L. (2015) *London Fog: The Biography*. Cambridge, MA: Harvard University Press.

Crate, S.A. and Nuttall, M. (eds) (2009) *Anthropology and Climate: From Encounters to Actions*. Walnut Creek, CA: Left Coast Press.

Cruikshank, J. (2001) 'Glaciers and climate change: Perspectives from oral tradition', *Arctic* 54 (4), 377–93.

Crutzen, P.J. and Stoermer, E.F. (2000) 'The "Anthropocene"', *IGBP Newsletter* 41: 17–18. http://www.igbp.net/download/18.316f18321323470177580001401/1376383088452/NL41.pdf (accessed 19 May 2016).

Culver, L. (2014) 'Seeing climate through culture', *Environmental History* 19 (2): 311–18.

Daston, L. (2010) 'The world in order', in D. Alberston and C. King (eds), *Without Nature? A New Condition for Theology*. Bronx, NY: Fordham University Press. pp. 15–34.

Davis, M. (2002) *Late Victorian Holocausts: El Niño Famines and the Making of the Third World*. London: Verso Books.

Donner, S. (2007) 'Domain of the gods: An editorial essay', *Climatic Change* 85 (3–4): 231–6.

Dove, M.R. (ed.) (2014) *The Anthropology of Climate Change: An Historical Reader*. Chichester: Wiley-Blackwell.

Dove, M. (2015) 'Historic decentering of the modern discourse of climate change: The long view from the Vedic Sages to Montesquieu', in J. Barnes and M. Dove (eds), *Climate Cultures: Anthropological Perspectives on Climate Change*. New Haven, CT: Yale University Press. pp. 24–57.

Doyle, J. (2007) 'Picturing the clima(c)tic: Greenpeace and the representational politics of climate change communication', *Science as Culture* 16 (2): 129–50.

Doyle, J. (2011) *Mediating Climate Change*. Farnham: Ashgate.

Draper, J.W. (1867) *History of the American Civil War*, 3 vols. New York: Harper.

Edwards, P.N. (2010) *A Vast Machine: Computer Models, Climate Data and the Politics of Global Warming*. Cambridge, MA: MIT Press.

Endfield, G.A. (2014) 'Exploring particularity: Vulnerability, resilience, and memory in climate change discourses', *Environmental History* 19: 303–10.

Ereaut, G. and Segnit, N. (2006) *Warm Words: How Are We Telling the Climate Story and Can We Tell It Better?* London: Institute for Public Policy Research. http://www.ippr.org/files/images/media/files/publication/2011/05/warm_words_1529.pdf (accessed 13 May 2016).

Fleming, J.R. (2010) *Fixing the Sky: The Checkered History of Weather and Climate Control*. New York: Columbia University Press.

Fleming, J.R. (2014) 'Climate physicians and surgeons', *Environmental History* 19: 338–45.

Fleming, J.R. and Janković, V. (eds) (2011) 'Klima', *Osiris* 26 (1) 1–226.

Flieger, V. and Anderson, D.A. (eds) (2014) *Tolkien on Fairy-Stories*. London: HarperCollins.

Fonstad, K.W. (1991) *The Atlas of Middle Earth*, revised edition. New York: Houghton Mifflin.

Fox, K. (2004) *Watching the English: the Hidden Rules of English Behaviour*. London: Hodder & Stoughton.

Garrard, G. (2013) 'Solar: Apocalypse not', in S. Goes (ed.), *Ian McEwan: Contemporary Critical Perspectives* (2nd edition). London: Bloomsbury Academic. pp. 123–36.

Geertz, C. (1973) *The Interpretation of Cultures: Selected Essays*. New York: Basic Books.

Glacken, C. (1967) *Traces on a Rhodian Shore: Nature and Culture in Western Thought from Ancient Times to the End of the Eighteenth Century*. Berkeley, CA: University of California Press.

Golinski, J. (2007) *British Weather and the Climate of Enlightenment*. Chicago, IL: Chicago University Press.

Gorman-Murray, A. (2010) 'An Australian feeling for snow: Towards understanding cultural and emotional dimensions of climate change', *Cultural Studies Review* 16 (1): 60–81.

Grattan, J. and Brayshay, M. (1995) 'An amazing and portentous summer: Environmental and social responses in Britain to the 1783 eruption of an Iceland volcano', *The Geographical Journal* 161 (2): 125–34.

Groom, N. (2013) *The Seasons: A Celebration of the English Year*. London: Atlantic Books.

Grove, R.H. and Adamson, G.C.D. (2017, in press) *El Niño in World History*. Palgrave London: MacMillan.

Gurevitch, L. (2014) 'Google warming: Google Earth as eco-machinima', *Convergence: The International Journal of Research into New Media Technologies* 20 (1): 85–107.

Hall, A. and Endfield, G. (2016) '"Snow scenes": Exploring the role of memory and place in commemorating extreme winters', *Weather, Climate, and Society* 8 (1): 5–19.

Hamilton, S. (2015) 'The global climate has always been broken: Failures of climate governance as global governmentality', *Caucasus International* 5 (2): 141–61.

Hammond, P. and Ortega Breton, H. (2014) 'Bridging the political deficit: Loss, morality and agency in films addressing climate change', *Communication, Culture and Critique* 7: 309–19.

Hare, F.K. (1966) 'The concept of climate', *Geography* 51: 99–110.

Hargreaves, J.C. and Annan, J.D. (2014) 'Can we trust climate models?', *WIREs Climate Change* 5 (4): 435–40.

Harrison, M. (1996) '"The tender frame of man": Disease, climate, and racial difference in India and the West Indies, 1760–1860', *Bulletin of the History of Medicine* 70: 68–93.

Harrison, S. (2004) 'Emotional climates: Ritual, seasonality and affective disorders', *Journal of the Royal Anthropological Institute* 10 (3): 583–602.

Hastrup, K.B. (2013) 'Anticipating nature: the productive uncertainty of climate models', in K. Hastrup and M. Skrydstrup (eds), *The Social Life of Climate Change Models: Anticipating Nature*. London and New York: Routledge. pp. 1–29.

Hau'ofa, E. (2008) *We are the Ocean: Selected Works*. Honolulu, HI: University of Hawaii Press.

Hauskeller, M. (2013) *Better Humans? Understanding the Enhancement Project*. Durham: Acumen.

Heise, U.K. (2008) *Sense of Place and Sense of Planet*. Oxford: Oxford University Press.

Hitchings, R. (2007) 'Geographies of embodied outdoor experience and the arrival of the patio heater', *Area* 39 (3): 340–8.

Hitchings, R. (2010) 'Seasonal climate change and the indoor city worker', *Transactions of the Institute of British Geographers* 35 (2): 282–98.

Höijer, B. (2010) 'Emotional anchoring and objectification in the media reporting on climate change', *Public Understanding of Science* 19 (6): 717–31.

Horn, R. (2007) *Weather Reports You*. Göttingen: Artangel/Steidl.

Huber, T. and Pedersen, P. (1997) 'Meteorological knowledge and environmental ideas in traditional and modern societies: The case of Tibet', *Journal of the Royal Anthropological Institute* 3 (3): 577–97

Hulme, M. (2008) 'The Conquering of climate: Discourses of fear and their dissolution', *The Geographical Journal* 174 (1): 5–16.

Hulme, M. (2009) *Why We Disagree About Climate Change: Understanding Controversy, Inaction and Opportunity*. Cambridge: Cambridge University Press.

Hulme, M. (2010) 'Problems with making and governing global kinds of knowledge', *Global Environmental Change* 20 (4): 558–64.

Hulme, M. (2011) 'Reducing the future to climate: A story of climate determinism and reductionism', *Osiris* 26 (1): 245–66.

Hulme, M. (2013) 'How climate models gain and exercise authority', in K. Hastrup and M. Skrydstrup (eds), *The Social Life of Climate Change Models: Anticipating Nature*. Abingdon: Routledge. pp. 30–44.

Hulme, M. (2014a) *Can Science Fix Climate Change? A Case Against Climate Engineering*. Cambridge: Polity Press.

Hulme, M. (2014b) 'Attributing weather extremes to "climate change": A review', *Progress in Physical Geography* 38 (4): 499–511.

Hulme, M. (ed.) (2015a) *Climates and Cultures: SAGE Library of the Environment*. 6 vols. London: Sage.

Hulme, M. (2015b) 'Climate and its changes: A cultural appraisal', *GEO: Geography and Environment* 2 (1): 1–11.

Hulme, M. (2015c) 'Better weather? The cultivation of the sky', *Cultural Anthropology* 30 (2): 236–44.

Hulme, M. (2016a) 'Climate change: Varieties of religious engagement', in W. Jenkins, M.E. Tucker and J. Grim (eds), *Routledge Handbook of Religion and Ecology*. Abingdon: Routledge. pp. 237–46.

Hulme, M. (2016b) 'Climate', in P. Whitfield (ed.), Part I. Mapping Shakespeare's World, in: *Volume 1: Shakespeare's World, 1500–1660*, in B.F. Smith (ed.) T*he Cambridge Guide to the Worlds of Shakespeare*. Cambridge: Cambridge University Press.

Hulme, M. (2016c) 'Climate change and memory', in S. Groes (ed.), *Memory in the Twenty-First Century: New Critical Perspectives From the Sciences, Arts and Humanities*. London: Palgrave Macmillan. pp. 159–62.

Huntington, E.W. (1915) *Civilization and Climate*. New Haven, MA: Yale University Press.

Ingold, T. (1994) 'Introduction to culture', in: T. Ingold, *Companion Encyclopedia of Anthropology: Humanity, Culture and Social Life*. London: Routledge. pp. 329–49.

Ingold, T. (2007) 'Earth, sky, wind, and weather', *Journal of the Royal Anthropological Institute (NS)* 13 (S1): S19–S38.

IPCC (2013) *Climate Change 2013: The Physical Science Basis. Contribution of Working Group I to the Fifth Assessment Report of the Intergovernmental Panel on Climate Change* [T.F. Stocker, D. Qin, G.-K. Plattner, M. Tignor, S.K. Allen, J. Boschung, A. Nauels, Y. Xia, V. Bex and P.M. Midgley (eds)]. Cambridge: Cambridge University Press.

Janković, V. (2006) 'Change in the weather', *Bookforum* (Feb/Mar): 39–41.

Janković, V. (2009) 'The end of weather: Outdoor garment industry and the quest for absolute comfort', in V. Janković,V. and C. Barboza (eds), *Weather, Local Knowledge and Everyday Life: Issues in Integrated Climate Studies*. Rio de Janeiro: MAST. pp. 172–180.

Janković, V. (2010) 'Climates as commodities: Jean Pierre Purry and the modelling of the best climate on Earth', *Studies in History and Philosophy of Science Part B: Studies in History and Philosophy of Modern Physics* 41 (3): 201–7.

Janković,V. (2014) 'Climate clichés: Overvaluation, fetishism and the ideologies of "national weather" in the long nineteenth century'. Paper presented at Birkbeck Forum, London, 12 March, https://www.academia.edu/6592112/Climate_clich%C3%A9s_overvaluation_fetishism_and_the_ideologies_of_national_weather_in_the_long_nineenth_century (accessed 9 September 2016).

Janković, V. (2015) 'The city', in K. Bäckstrand and E. Lövbrand (eds), *Research Handbook on Climate Governance*. Cheltenham: Edward Elgar Publishing. pp. 332–342.

Jenkins, W. (2005) 'Assessing metaphors of agency: Intervention, perfection and care as models of environmental practice', *Environmental Ethics* 27 (Summer): 135–54.

Jenkins, W. (2016) 'Stewards of irony: Planetary stewardship, climate engineering and religious ethics', in (Chapter 8) F. Clingerman and K.J. O'Brien (eds), *Calming the Storm: Theological and Ethical Perspective on Climate Engineering*. Lanham, MD: Lexington Books.

Johns-Putra, A. (2016) 'Climate change in literature and literary studies: From cli-fi, climate change theater and ecopoetry to ecocriticism and climate change criticism', *WIREs Climate Change* 7 (2): 266–82.

Johnson, J. (1818) *The Influence of Tropical Climates on European Constitutions* (2nd edition). London.

Judd, J.W. (2015) 'Has "Game of Thrones" affected the way people think about climate change?' https://psmag.com/has-game-of-thrones-affected-the-way-people-think-about-climate-change-dcafed0fe5e7#.kdqu5ho06 (accessed 19 May 2016).

Jung, K., Shavitt, S., Viswanathan, M. and Hilbe, J.M. (2014) 'Female hurricanes are deadlier than male hurricanes', *Proceedings of the National Academy of Sciences* 111 (24): 8782–7.

Keith, D. (2013) *A Case for Climate Engineering*. Cambridge, MA: MIT Press.

Kellogg, W.W. and Schneider, S.H. (1974) 'Climate stabilization: For better or for worse?' *Science* 186 (4170): 1163–72.

Kempf, W. (2015) 'Representation as disaster: Mapping islands, climate change and displacement in Oceania', *Pacific Studies* 38 (1–2): 200–28.

Kennedy, J.F.K. (1963) 'Address at the Anniversary Convocation of the National Academy of Sciences October 22, 1963', http://www.presidency.ucsb.edu/ws/?pid=9488 (accessed 19 May 2016).

Klein, N. (2014) *This Changes Everything: Capitalism vs. the Climate*. New York: Simon & Schuster.

Klinenberg, E. (2002) *Heatwave: A Social Autopsy of Disaster in Chicago*. Chicago, IL: University of Chicago Press.

Knebusch, J. (2008) 'Art and climate (change) perception: Outline of a phenomenology of climate change', in S. Kagan and V. Kirchberg (eds), *Sustainability: A New Frontier for the Arts and Cultures*. Frankfurt: Verlag für Akademische Schriften. pp. 242–62.

Lahsen, M. (2005) 'Seductive simulations? Uncertainty distribution around climate models', *Social Studies of Science* 35: 895–922.

Lamb, H.H. (1959) 'Our changing climate, past and present', *Weather* 14, 299–318.

Latour, B. (1993) *We Have Never Been Modern* (trans. C. Porter). New York: Harvester/Wheatsheaf.

Latour, B. (2010) 'An attempt at a "Compositonist manifesto"', *New Literary History* 41: 471–90.

Latour, B. (2014) 'Agency at the time of the Anthropocene', *New Literary History* 45, 1–18. http://www.bruno-latour.fr/sites/default/files/128-FEL-SKI-HOLBERG-NLH-FINAL.pdf (accessed 11 May 2016).

Leduc, T.B. (2007) 'Sila dialogues on climate change: Inuit wisdom for a cross-cultural inter-disciplinarity', *Climatic Change* 85 (3/4): 237–50.

Leduc, T.B. (2010) *Climate, Culture, Change: Inuit and Western Dialogues With a Warming North*. Ottawa: University of Ottawa Press.

Leduc, T.B. (2016) *A Canadian Climate of Mind: Passages From Fur to Energy and Beyond*. Montreal: McGill-Queen's University Press.

Lehmann, P.N. (2016) 'Infinite power to change the world: Hydroelectricity and engineered climate change in the Atlantropa Project', *American Historical Review* 121 (1): 70–100.

Lemos, M.C. and Rood, R.B. (2010) 'Climate projections and their impact on policy and practice', *WIREs Climate Change* 1 (5), 670–82.

Lie, J. (2007) 'Global climate change and the politics of disaster', *Sustainability Science* 2: 233–6.

Livingstone, D.N. (1999) 'Tropical climate and moral hygiene: The anatomy of a Victorian debate', *British Journal for the History of Science* 32 (1): 93–110.

Livingstone, D.N. (2004) 'Climate', in N. Thrift, S. Harrison and S. Pile (eds), *Patterned Ground: Entanglements of Nature and Culture*. London: Reaktion Books. pp. 77–9.

Livingstone, D.N. (2015) 'The climate of war: violence, warfare, and climatic reductionism', *WIREs Climate Change* 6 (5): 437–44.

Locher, F. and Fressoz, J.B. (2012) 'The frail climate of modernity. A climate history of environmental reflexivity', *Critical Inquiry* 38 (3): 579–98.

Lovelock, J. (2006) 'The earth is about to catch a morbid fever that may last as long as 100,000 years', *The Independent*, 16 January. http://www.independent.co.uk/voices/commentators/james-lovelock-the-earth-is-about-to-catch-a-morbid-fever-that-may-last-as-long-as-100000-years-5336856.html (accessed 13 May 2016).

Lüdecke, C. (2009) '"I always feel the föhn, even if it's not there". The Bavarian föhn phenomenon in everyday life', in V. Janković and C. Barboza (eds), *Weather, Local Knowledge and Everyday Life: Issues in Integrated Climate Studies*, Rio de Janeiro: MAST. pp. 209–18.

Lunt, D.J. ['Radagast the Brown'] (2013) 'The climate of Middle Earth', http://www.bristol.ac.uk/university/media/press/10013-english.pdf (accessed 16 May 2016).

Lynas, Mark (2007) 'Global warming: The final warning', *The Independent*, 3 February, http://terranature.org/IPCCwarning2007.htm (accessed 20 May 2016).

Mahony, M. (2016) 'For an empire of "all types of climate": Meteorology as an imperial science', *Journal of Historical Geography* 51: 29–39.

Mahony, M. and Hulme, M. (2016) Modelling and the nation: Institutionalizing climate prediction in the UK, 1988–92, *Minerva* online at doi: 10.1007/s11024-016-9302-0.

Marris, E. (2011) *Rumbunctious Garden: Saving Nature in a Post-Wild World*. London: Bloomsbury.

Martin, C. (2006) 'Experience of the New World and Aristotelian revisions of the Earth's climates during the Renaissance', *History of Meteorology* 3: 1–16.

Mathur, N. (2015) '"It's a conspiracy theory and climate change". Of beastly encounters and cervine disappearances in Himalayan India', *HAU: Journal of Ethnographic Theory* 5 (1): 87–111.

McCusker, K.E., Battisti, D.S. and Bitz, C.M. (2012) 'The climate response to stratospheric sulfate injections and implications for addressing climate emergencies', *Journal of Climate* 25: 3096–3116.

McIntosh, R.J. (2015) 'Climate shock and awe: Can there be an "ethno-science" of deep-time Mande paleoclimate memory?' in J. Barnes and M.R. Dove (eds), *Climate Cultures: Anthropological Perspectives on Climate Change*. New Haven, CT: Yale University Press. pp. 273–88.

Meyer, W.B. (2000) *Americans and Their Weather*. Oxford: Oxford University Press.

Meyer, W.B. (2002) 'The perfectionists and the weather: The Oneida Community's quest for meteorological utopia 1848–1879', *Environmental History* 7 (4): 589–610.

Meze-Hausken, E. (2007) 'Seasons in the sun – weather and climate front-page news stories in Europe's rainiest city, Bergen, Norway', *International Journal of Biometeorology* 52 (1): 17–31.

Miles, M. (2010) 'Representing nature: Art and climate change', *Cultural Geographies* 17 (1): 19–35.

Milkoreit, M. (2016) 'The promise of climate fiction: imagination, storytelling and the politics of the future', in P. Wapner and E. Hilal (eds), *Reimagining Climate Change*. Abingdon: Routledge. pp. 171–91.

Miller, P. and Rose, N. (1990) 'Governing economic life', *Economy and Society* 19 (1): 1–31.

Moore, P. (2015) *The Weather Experiment: The Pioneers Who Sought to See the Future*. London: Chatto & Windus.

More, M. (2013) 'Letter to Mother Nature', in M. More and N. Vita-More (eds), *The Transhumanist Reader: Classical and Contemporary Essays on the Science, Technology, and Philosophy of the Human Future*. Oxford: Wiley Blackwell. pp. 449–450.

NDU (1978) *Climate Change to the Year 2000: A Survey of Expert Opinion*. Fort Lesley J McNair. Washington, DC: National Defense University. Available from http://eric.ed.gov/?id=ED160394 (accessed 16 May 2016).

Nerlich, B. and Jaspal. R. (2012) 'Metaphors we die by? Geoengineering, metaphors and the argument from catastrophe', *Metaphor and Symbol* 27 (2): 131–47.

Neuberger. H. (1970) 'Climate in art', *Weather* 25: 46–56.

Neumann, J. von (1955) 'Can we survive technology?', *Fortune* 91 (6): 32–47.

Northcott, M.S. (2007) *Moral Climate: The Ethics of Global Warming*. London: Dartman, Longman and Todd Ltd/Orbis.

O'Brien, K. (2013) 'Global environmental change III: Closing the gap between knowledge and action', *Progress in Human Geography* 37 (4): 587–96.

O'Reilly, J. (2015) 'Glacial dramas: Typos, projections and peer review in the Fourth Assessment of the Intergovernmental Panel on Climate Change', in J. Barnes and M. Dove (eds), *Climate Cultures: Anthropological Perspectives on Climate Change.* New Haven, CT: Yale University Press. pp. 107–26.

Oreskes, N. and Conway, E.M. (2014) *The Collapse of Western Civilisation: A View from the Future.* New York: Columbia University Press.

Oreskes, N., Shrader-Frechette, K. and Belitz, K. (1994) 'Verification, validation and confirmation of numerical models in the earth sciences', *Science* 263: 641–6.

Orlove, B., Roncoli, C., Kabugo, M. and Majigu, A. (2010) 'Indigenous climate knowledge in southern Uganda: The multiple components of a dynamic regional system', *Climatic Change* 100 (2): 243–65.

Paerregaard, K. (2013) 'Bare rocks and fallen angels: Environmental change, climate perceptions and ritual practice in the Peruvian Andes', *Religions* 4 (2): 290–305.

Pal, J.S. and Eltahir, E.A.B. (2016) 'Future temperature in southwest Asia projected to exceed a threshold for human adaptability', *Nature Climate Change* 6 (2): 197–200.

Paola, M. di (2015) 'Virtues for the Anthropocene', *Environmental Values* 24: 183–207.

Parker, G. (2013) *Global Crisis: War, Climate Change and Catastrophe in the Seventeenth Century.* New Haven, CT: Yale University Press.

Perier, J.-N. (1845) 'De l'acclimatement en Algérie', *Annales d'Hygiène Publique et de Médecine Légale* 33: 40–41.

Pleij, H. (2001) *Dreaming of Cockaigne: Medieval Fantasies of the Perfect Life,* trans. D. Webb. New York: Columbia University Press.

Plumwood, V. (1993) *Feminism and the Mastery of Nature.* Abingdon: Routledge.

Porter, P.W. and Lukermann, F.E. (1976) 'The geography of utopia', in D. Lowenthal and M.J. Bowden (eds), *Geographies of the Mind: Essays in Historical Geosophy in Honour of John Kirtland Wright.* New York: Oxford University Press. pp. 197–204.

Powell, R.C. (2015) 'History and philosophy of geography II: The future history of the geographical propaedeutic?', *Progress in Human Geography* 39 (4): 486–96.

Randall, R. (2009) 'Loss and climate change: The cost of parallel narratives', *Ecopsychology* 1 (3): 118–29.

Rayner, S. (2003) 'Domesticating nature: Commentary on the anthropological study of weather and climate discourse', in S. Strauss and B. Orlove (eds), *Weather, Climate, Culture.* Oxford: Berg. pp. 277–90.

Rice, J.L., Burke, B.J. and Heynen, N. (2015) 'Knowing climate change, embodying climate praxis: Experiential knowledge in Southern Appalachia', *Annals of the Association of American Geographers* 105 (2): 253–62.

Rockoff, M. and Meisch, S. (2015) 'Climate change in early modern literature: Which place for humanities in the sustainability sciences?' in S. Meisch,

J. Lundershausen, L. Bosser and M. Rockoff (eds), *Ethics of Science in the Research for Sustainable Development*. Baden-Baden: Nomos. pp. 269–98.

Romm, J. (2015) 'Meet the world's expert on climate change and "Game Of Thrones"'. http://thinkprogress.org/climate/2015/04/10/3645070/climate-change-game-thrones/ (accessed 2 May 2016).

Roncoli, C., Ingram, K., Jost, C. and Kirshen, P. (2003) 'Meteorological meanings: Farmers' interpretations of seasonal rainfall forecasts in Burkina Faso', in S. Strauss and B. Orlove (eds), *Weather, Climate, Culture*. Oxford: Berg. pp. 181–202.

Rosteck, T. and Frentz, T.S. (2009) 'Myth and multiple readings in environmental rhetoric: The case of *An Inconvenient Truth*', *Quarterly Journal of Speech* 95 (1): 1–19.

Roudaire, F.E. (1874) 'Une Mer intérieure en Algérie', *Revue des Deux Mondes* 3: 323–50.

Rubin, J.H. (1996) 'Realism', in J. Turner (ed.), *The Dictionary of Art* (vol. 26), London: Grove Publishers. pp. 52–7.

Rudiak-Gould, P. (2012) 'Promiscuous corroboration and climate change translation: A case study from the Marshall Islands', *Global Environmental Change* 22 (1): 46–54.

Rudiak-Gould, P. (2013) '"We have seen it with our own eyes": Why we disagree about climate change visibility', *Weather, Climate and Society* 5 (2): 120–132.

Sachs, J.D., Mellinger, A.D. and Gallup, J.L. (2001) 'The geography of poverty and wealth', *Scientific American* (3): 71–4.

Salvador, M. and Norton, T. (2011) 'The flood myth in the age of global climate change', *Environmental Communication* 5 (1): 45–61.

Sanders, T. (2014) 'Comment on "Climate change and accusation: Global warming and local blame in a small island state" by Peter Rudiak-Gould', *Current Anthropology* 55 (4): 381–382.

Schneider, B. (2012) 'Climate model simulation visualisation from a visual studies perspective', *WIREs Climate Change* 3 (2): 185–93.

Scott, J.C. (1998) *Seeing Like a State: How Certain Schemes to Improve the Human Condition Have Failed*. New Haven, CT: Yale University Press.

Sen, A. (1981) *Poverty And Famines: An Essay on Entitlement and Deprivation*. Oxford: Oxford University Press.

Shapiro, J. (2009) *1599: A Year in the Life of William Shakespeare*. London: Faber & Faber.

Shove, E. (2003) 'Converging conventions of comfort, cleanliness and convenience', *Journal of Consumer Policy* 26: 395–418.

Sillmann, J., Lenton, T., Levermann, A., Ott, K., Hulme, M., Benduhn, F. and Horton, J.B. (2015) 'Climate emergency – no argument for climate engineering', *Nature Climate Change* 5 (4): 290–2.

Simpson, D. (1982) *Fetishism and Imagination: Dickens, Melville, Conrad*. Baltimore, MD: Johns Hopkins University Press.

Skrimshire, S. (2014) 'Climate change and apocalyptic faith', *WIREs Climate Change* 5 (2): 233–46.

Smith, P. and Howe, N. (2015) *Climate Change as Social Drama: Global Warming in the Public Sphere.* New York: Cambridge University Press.

Solli, J. and Ryghaug, M. (2014) 'Assembling climate knowledge: The role of local expertise', *Nordic Journal of Science and Technology Studies* 2 (2): 18–28.

Sontag, S. (1977) *On Photography.* New York: Farrar, Straus & Giroux.

Sörlin, S. (2011) 'The anxieties of a science diplomat: Field coproduction of climate knowledge and the rise and fall of Hans Ahlman's "Polar Warming"', *Osiris* 26 (1): 66–88.

Stehr, N. (1997) 'Trust and climate', *Climate Research* 8 (3): 163–9.

Stevens, W.K. (1998) 'If climate changes it may change quickly', *New York Times*, 27 January. http://www.nytimes.com/1998/01/27/science/if-climate-changes-it-may-change-quickly.html?pagewanted=all (accessed 11 May 2016).

Storch, H. von and Stehr, N. (2006) 'Anthropogenic climate change: A reason for concern since the 18th century and earlier', *Geografiska Annaler* 88A (2): 107–13.

Strauss, S. and Orlove, B. (eds) (2003) *Weather, Climate, Culture.* Oxford/New York: Berg.

Strengers, Y. and Maller, C. (2013) Getting comfortable: the materiality of weather in household heating and cooling practices Unpublished conference presentation at 'Reflections, intersections and aspirations: 50 years of Australian sociology' The Australian Sociological Association Conference, Monash University, 25–28 November.

Svoboda, M. (2016) 'Cli-fi on the screen(s): Patterns in the representations of climate change in fictional films', *WIREs Climate Change* 7 (1): 43–64.

Szerszynski, B. (2010) 'Reading and writing the weather: Climate technics and the moment of responsibility', *Theory, Culture & Society* 27 (2/3): 9–30.

Tambe, A. (2011) 'Climate, race science and the age of consent in the League of Nations', *Theory, Culture & Society* 28 (2): 109–30.

Thornes, J.E. and McGregor, G.R. (2003) 'Cultural climatology', in S. Trudgill and A. Roy (eds), *Contemporary Meanings in Physical Geography: From What to Why.* London: Routledge. pp. 173–97.

Thornes, J.E. and Metherill, G. (2003) 'Monet's London series and the cultural climate of London at the turn of the twentieth century', in S. Strauss and B. Orlove (eds), *Weather, Climate, Culture.* Oxford: Berg. pp. 141–60.

Trewartha, G.T. (1961) *The Earth's Problem Climates.* Madison, WI: University of Wisconsin Press.

Trexler, A. and Johns-Putra, A. (2011) 'Climate change in literature and literary criticism', *WIREs Climate Change* 2 (2): 185–200.

Tyszczuk, R. (2014) 'Cautionary tales: The sky is falling! The world is ending!' in J. Smith, R. Tyszczuk and R. Butler, *Culture and Climate Change: Narratives.* Cambridge: Shed Publishing. pp. 45–57.

Vallisneri, A. (1721) *Dei corpi marini che sui monti si trovano [Of Marine Bodies Found in the Mountains].* Venice. Available from https://archive.org/details/bub_gb_V73O4lmASCkC (accessed 9 May 2016).

Vannini, P., Waskul, D., Gottschalk, S. and Ellis-Newstead, T. (2012) 'Making sense of the weather: Dwelling and weathering on Canada's rain coast', *Space and Culture*, 15 (4): 361–380.

Vet, E. de (2013) 'Exploring weather-related experiences and practices: examining methodological approaches', *Area* 45 (2): 198–206.

Vet, E. de (2014) 'Weather-ways: Experiencing and responding to everyday weather', PhD thesis, Department of Geography and Sustainable Communities, University of Wollongong, Australia. http://ro.uow.edu.au/theses/4244 (accessed 2 May 2016).

Walsh, L. (2013) *Scientists as Prophets: A Rhetorical Genealogy.* New York: Oxford University Press.

Walsh, L. (2015) 'The visual rhetoric of climate change', *WIREs Climate Change* 6 (4): 361–8.

Watsuji, T. (1988[1935]) *Climate and Culture: A Philosophical Study* (trans. G. Bownas). New York: Greenwood Press.

Watt, N. (2006) 'Blair warns of climate change "tipping points"', *The Guardian.* http://www.theguardian.com/world/2006/oct/20/greenpolitics.politics (accessed 11 May 2016).

Welzer, H., Soeffner, H.-G. and Giesecke, D. (eds) (2010) *KlimaKulturen: Soziale Wirklichkeiten im Klimawandel.* Frankfurt: Campus Verlag.

Wilson, A. (1992) *The Culture of Nature: North American Landscape from Disney to the Exxon Valdez.* Oxford: Blackwell.

Wilson, E.K. (2012) *After Secularism: Rethinking Religion in Global Politics.* Basingstoke: Palgrave.

World Meteorological Organisation (WMO) (n.d.) 'Frequently Asked Questions (FAQs)'. http://www.wmo.int/pages/prog/wcp/ccl/faqs.php (accessed 2 May 2016).

# Index

Note: The index covers the main text but not the bibliography. The suffixes f and n after a page number indicate that that the topic is covered only in an illustration or footnote on that page.